Mathematics in Ancient Jaina Literature

Mathematics in Ancient Jaina Literature

Edited by

S. G. Dani
Centre for Excellence in Basic Sciences, Mumbai, India

S. K. Jain
Ohio University, USA

With the co-operation of **Pankaj K. Shah**
Jain Center of Greater Boston, MA, USA

World Scientific

NEW JERSEY • LONDON • SINGAPORE • BEIJING • SHANGHAI • HONG KONG • TAIPEI • CHENNAI • TOKYO

Published by

World Scientific Publishing Co. Pte. Ltd.
5 Toh Tuck Link, Singapore 596224
USA office: 27 Warren Street, Suite 401-402, Hackensack, NJ 07601
UK office: 57 Shelton Street, Covent Garden, London WC2H 9HE

Library of Congress Control Number: 2022032485

British Library Cataloguing-in-Publication Data
A catalogue record for this book is available from the British Library.

MATHEMATICS IN ANCIENT JAINA LITERATURE

ISBN 978-981-125-549-6 (hardcover)
ISBN 978-981-125-550-2 (ebook for institutions)
ISBN 978-981-125-551-9 (ebook for individuals)

For any available supplementary material, please visit
https://www.worldscientific.com/worldscibooks/10.1142/12813#t=suppl

Typeset by Stallion Press
Email: enquiries@stallionpress.com

Printed in Singapore

https://doi.org/10.1142/9789811255502_fmatter

Preface

This monograph is an outgrowth of an international ZOOM conference held in December 2020 on the history of Mathematics in ancient Jain literature, which was originally scheduled as a conference in conventional form at MIT Boston in September 2020, but had to be cancelled on account of the pandemic. This was sponsored by the Jain Center of Greater Boston and the Academic Liaison Committee of the Federation of Jain Associations in North America. There were 25 speakers and over two hundred participants on ZOOM from across the world, including, notably, Fields Medalists Dr. Efim Zelmanov (UCSD) and Dr. Manjul Bhargava (Princeton).

The main objective before the speakers was to bring to the attention of historians in mathematics that there is a plenty of literature written by monks and scholars in the Jain works that contains elements of arithmetic, algebra and geometry, independent of discoveries by other cultures in the past. The talks and the discussions at the conference highlighted a need for a monograph that can be recommended as a reference book for a course on History of Mathematics in the Departments of Mathematics and Education in colleges and universities. This is our hope that the present volume would fill the gap in the knowledge of past Jain contributions.

Besides the support by the Jain Center of Greater Boston, the Sir John Templeton Foundation and the Uberoi Foundation, USA, provided major financial support. We thank all these agencies for their generous support.

Several individuals have made this project a success. These include Mr. Pankaj Shah, Dr. Sulekh Jain, Dr. Manoj Jain, Dr. Anupam Jain and Dr. Devavrat Shah, for their vision to initiate and organize this conference at MIT. In particular, our hearts go to Mr. Pankaj Shah who took upon himself the responsibility with great passion to ensure that everything was going well in organizing the conference and that the speakers submitted their talks and papers in a timely fashion.

The publication of this monograph could not have been possible without the arduous support of several individuals. We would like to mention, in particular, Dr. R. N. Gupta (Chandigarh, India), Dr. Ravinder Kumar (Cumming, GA, USA), Dr. R. C. Gupta (Jhansi, India), Dr. K. M. Prasad (Manipal, India), Dr. J. Radhakrishnan (Mumbai, India), Dr. Raj Kumar Jain (St Louis, MO, USA), Dr. Vijay Vaishnavi (Cary, NC, USA), Dr. Vinod Jain (Dayton, OH, USA), Dr. S. A. Katre (Pune, India), Mr. S. M. Jain (Phoenix, AZ, USA), Dr. Jitender Jadhav (Barwani, India), Dr. Martin Mohlenkamp (Athens, OH, USA), Dr. Steven R. Jain (Dayton, OH, USA), Mr. Umashankara Kelathaya (Manipal, India), and Mr. Ratnakumar S. Shah (Pune, India) for providing their professional services in several ways that led to the final outcome of this manuscript for publication. Our special thanks to Dr. Vinod Jain (Dayton, OH, USA) for putting in immense energy in sorting the papers with respect to quality. We would also like to thank Dr. S. C. Agrawal (Meerut, UP, India) for giving pieces of professional advice on some of the papers. Last but not least, we thank the authors for their excellent contributions.

We would also like to thank the publication editor Ms. Rok Ting Tan of the World Scientific Publishing Company for her help throughout the process.

S. G. Dani and S. K. Jain

https://doi.org/10.1142/9789811255502_fmatter

Introduction

This volume is a collection of invited articles in the history of mathematics. The volume surveys the mathematical contents and knowledge found in ancient Jaina literature. There are very limited resources that are currently available on this topic. Reliable presentation is available only in a few short accounts, which are isolated and rather inaccessible. Our goal in bringing out this monograph has been to produce a compendium of articles giving a reasonably comprehensive account of the mathematical ideas and achievements of the Jaina scholars. The articles are aimed at a broader audience. We have also endeavored to meet appropriate scholarly standards, and to adhere to the contemporary global norms in the study of history of mathematics.

The first article, by R.C. Gupta, provides a comprehensive history of decimals and zeros in the Jaina literature. This is followed by two articles with detailed accounts of concepts and ideas in ancient and medieval Jaina works, one on arithmetic and combinatorics by Ratnakumar S. Shah, and another on geometry by S.G. Dani. The remaining nine articles are on specific themes. Among these, apart from those on topics in arithmetic, combinatorics and geometry, there are two articles connecting to contemporary pursuits, one involving an instance of use of cryptography, and another dealing with Jaina logic and its application to statistical thinking.

Here is a brief description of the contents of the monograph, together with our perspective on the articles.

The author of the opening article is R.C. Gupta, a renowned historian of mathematics and a winner of the coveted Kenneth O. May prize in the subject, awarded by the International Commission on History of Mathematics. He has contributed an exhaustive article, "Decimal numerals and zero in Jaina literature". It is well known that ancient India has had a major role in the evolution of decimal representation of numbers, together with zero, which has been very impactful in the development of mathematics worldwide. The ancient Jaina scholars are noted to have been deeply involved with various key features driving this development, including usage of large numbers, decimal representations with decuple terms going up to very high denominations, the very enigmatic zero, and the associated arithmetic. In this article the author meticulously traces the progress along this broad theme in ancient Jaina literature. He also dispels along the way various misconceptions that have been in circulation.

The contemplation of the universe by Jaina scholars, and the elaborate models they conceived for the structure of the universe and the human population, led them to formulate, and to deliberate on, a variety of mathematical concepts. Many notions and ideas they introduced bear interesting kinship with various concepts in modern mathematics. These relate to a broad spectrum of topics, starting with generalities in arithmetic, combinatorics and geometry, extending to principles of algebra, and also incorporating some advanced topics like probability, transfinite numbers, and logarithms. In certain cases the small beginnings made in the ancient canonical literature served as a basis for much advance on the topics by later adherents, especially Vīrasena and Nemicandra, near the end of the first millennium. Article 2, "Evolution and development of ideas through Jaina scriptures" by Ratnakumar S. Shah, gives a detailed account in this respect. This account is based on the canonical works from the 4th century BCE to 5th century CE, and various commentaries from the later period until the 15th century.

The cosmological model of the universe, and the "geographical" details of the Jambudvīpa (the earth) that the Jaina scholars envisioned, evidently inspired them into thinking about a variety of questions in geometry. Unfortunately, the extant works do not provide us adequate insight into their thinking on the subject. However, various canonical compositions are found to contain interesting details pertaining to specific geometrical relations; the latter were presumably limited to what fascinated the scholars. Additionally, a variety of calculations concerning astral bodies, in the form they were thought of, also throw light on their familiarity with various geometrical principles. Along with what is normally part of geometry, one also finds here some interesting (approximate) formulae interrelating lengths of arcs and chords of circles, which may be viewed as a sort of proto-trigonometry. These ideas are also seen to be carried forward, and further developed by the scholars during the revived phase of Jaina scholarship in the 8th and 9th centuries. In particular, some interesting new applications of the ancient ideas are found in *Gaṇita sāra saṅgraha* of Mahāvīra. A discussion on these geometric contents is the subject of article 3, "Geometry in ancient Jaina works: a review" by S. G. Dani.

Following these articles, which relate to the development of ideas in the Jaina tradition over extended periods, there are articles focusing on various specific themes that engaged the Jaina scholars in the course of their pursuits.

Article 4 by Dipak Jadhav is on the count of **partitions** of natural numbers. In modern mathematics such a count, encoded in what is called the partition function, plays an important role. One may recall here that a celebrated result of G.H. Hardy and S. Ramanujan gives an asymptotic formula for the partition function. It is interesting to note, following this article, that the idea of keeping such a count goes back to the ancient work *Bhagavatī sūtra* (codified in 466 CE), where partitions have been elaborately listed, and counts described, for numbers upto 10, and also discussed in the Jaina framework of large numbers. The theme, however, does not seem to have been pursued later in classical Indian mathematics.

Article 5 by Jitendra K. Sharma explores the Jaina thought on the theme of **infinity**. The idea of infinity has fascinated the Indian mind down the ages. In the Jaina scheme of things, going beyond the general awe about it, one also sees an attempt to classify infinities, and to place the idea, so to speak, in a broader perspective on the numeration or quantification of entities. While it is not clear if the notions that emerged through their thought process can be placed in a rigorous mathematical framework of our times, the general similarity in the approach seems worth noting, as a matter of historical record.

Articles 6 and 7 discuss two special topics contained in the classic work *Gaṇita sāra saṅgraha* (GSS), composed by the renowned 9th century Jaina mathematician Mahāvīra. The work represents a mature phase in Indian mathematics, in terms of pedagogy and dissemination of mathematics in the society at large. It is a comprehensive composition with over 1100 verses, divided into nine chapters, incorporating descriptions of various mathematical principles and algorithms, followed by numerous illustrative examples. Evidence suggests that it served as a textbook of mathematics over a wide region for several centuries.

In article 6, S. Balachandra Rao, K. Rupa, S.K. Uma, and Padmaja Venugopal discuss the chapter in *Gaṇita sāra saṅgraha* (GSS) devoted to *kṣetravyavahāra*, viz. **mensuration of planar figures**, as elucidated by Mahāvīra. That chapter is the second largest of the 9 chapters in the work, with well over 200 verses. Apart from the usual basic geometric figures, GSS discusses in this chapter mensuration of various non-standard figures, e.g. the shape of a conch, viewing them suitably as composites of basic figures. There is also a discussion on mensuration of a geometric figure called "āyatavṛtta" ("oblong circle"), which is rather unique to GSS. The authors of this article give a detailed exposition of the formulae described in the chapter, along with some examples. Analogous formulae given in other ancient Indian works are also recalled at various points for comparison.

The **indeterminate linear equations**, known in Jaina literature as **Kuṭṭīkāra**, is a familiar topic in ancient Indian works starting with Āryabhaṭa's *Āryabhaṭīya* (499 CE). It concerns finding integer solutions x, y of an equation of the form $ax - by = c$, where a and b are positive integers and c is an integer (positive or negative). Such problems arise naturally in mathematical astronomy, and also in various practical contexts. At its core, the approach introduced by

Āryabhaṭa, consists of simplifying the equation successively (in terms of the size of the coefficients) by a process of mutual division, until one is able to solve the equation; such a stage necessarily arises when c is divisible by the greatest common divisor of a and b. There are various matters of detail on how to organize the process, for efficient execution, which seem to have evolved over the centuries following *Āryabhaṭīya*. In article 7, Sudarshan B. and N. Shivakumar give an exposition of the method as described in GSS, together with various examples.

Formulae for the **summation of series** (with finitely many terms) has been a topic considered with much interest in Indian mathematics. The Jaina scholars, especially of the later phase, have worked extensively on this topic. Article 8 by R.C. Gupta discusses summation formulae for a variety of series based on, or derived from the arithmetic series, considered by Śrīdhara (mid 8th century) and Mahāvīra (mid 9th century), and puts them in a common framework by introducing what he calls Śrīdhara-Mahāvīra series.

There is an interesting case of a text from the 9th century, *Siribhūvalaya*, written by a Jaina scholar named Kumudendu, involving encryption and decryption methods in the sense of **cryptography**. Article 9, by Anil Kumar Jain, gives a description of the text and its methods in which cryptographic notions and techniques are involved in the process, including features for error correction akin to parity-check bits in error-correcting codes.

Logic is the thread that links philosophy to mathematics. No wonder, therefore, that the mathematically-minded Jaina philosophers also developed principles of logic. The core concept of Jaina logic, originating from ancient Jaina literature, consists of a conditional holistic principle, which has been studied in modern terminology by statisticians. Article 10 by Kanti V. Mardia and Anthony J. Ruda analyzes aspects of ancient Jaina logic and draws connections to present-day statistical thinking which could be of interest especially in Bayesian analysis and interdisciplinary data science.

Article 11 by Achary Vijay Nandighoshsuri recounts the **transition from oral tradition** in preservation of scriptures to recording them in the palm-leaf medium, and infers that the use of the decimal system and zero in Jaina tradition goes back to the late 4th or early 3rd century BCE.

It is evident that in the overall context, a lot more exploration is needed into the mathematics of the ancient Jainas. This requires looking into ancient texts in various repositories, and also perhaps some unexpected places. A good deal of work would also be needed in editing them so as to make the relevant contents accessible to scholars for further analysis, and to the wider audience. There is an ongoing effort in this context to keep track of the **extant manuscripts**, in which Anupam Jain, the author of the concluding article, has been actively involved. In his article he provides information on available manuscripts, going back to the 8th century, together with information on the status regarding the publication of the texts.

We conclude this introduction with a hope that this publication will inspire and lead to future researches in understanding and exploring hidden treasures in ancient Jaina works and relating them to modern mathematical concepts.

— *Editors*

https://doi.org/10.1142/9789811255502_fmatter

Contents

https://doi.org/10.1142/9789811255502_0001

Article 1

Decimal Numerals and Zero in Ancient Jaina Literature

R. C. Gupta

R-20, Ras Bahar Colony, P.O. Sipri Bazar
Jhansi 284003, U.P., India

The most ancient literature of the Jainas consists of the *Āgamas* (Canons) of various categories such as *Aṅgas*, *Upāṅgas*, etc. Also, there are other sacred works such as *Ṣaṭkhaṇḍāgama* and *Kasāyapāhuḍa* etc. In ancient times numbers were written in words and counting was in decimal scale. In early times the name *koṭi* (ten million) was repeatedly used for large numbers but regular decimal place-value system (*daśaguṇottara*) names to 18 and higher order are found in the works of Śrīdhara and Mahāvīrācārya. A special time period of 8400,000 years gave rise to a new scale for reckoning time upto *Śirṣaprahelikā*. The decimal place-value method of counting was clearly known to the author. The *Anuyoga-dvāra-sūtra* which correctly states that a very large specific number occupies 29 places in the decimal scale. The famous Prakrit work *Tiloya-paṇṇattī* very explicitly uses with ease the modern place-value system of numerals with zero. This means that by the fifth century CE the system was well known to Jainas and implies its knowledge to Jaina mathematicians still earlier. Rules for operations of addition, subtraction, multiplication and division (to some extent) involving zero are found in clear terms in Śrīdhara and Mahāvīrācārya who made full use of the decimal place-value system with *bhūta-saṁkhyas* (with zero).

Keywords: Ancient and medieval Jaina mathematics, Śrīdhara, Mahāvīrācārya, decimal place-value counting and numerals, zero.

Mathematics Subject Classification 2020: 01-06, 01A32, 01A99.

1.1. Introduction

Jainism is one of the most ancient religions of the world. Jainas have contributed significantly in many areas of knowledge such as logic, philosophy, spiritual and secular sciences, and mathematics (*gaṇita*).

The Jaina canonical literature (*āgama*) is divided into various categories such as *Aṅgas, Upāṅgas, Cūlikās*, etc. The *Aṅgas* are twelve and their names are as follows [1]:

(1) Āyāraṃga, (2) *Sūyagadaṃga*, (3) *Ṭhāṇāṃga*, (4) *Samavāyāṃga*, (5) *Viyāha-paṇṇatti* (or *Bhagavatī*), (6) *Nāyādhamma-kahāo*, (7) *Upāsaga-dasāo*, (8) *Aṃtagaḍa-dasāo*, (9) *Aṇuttarova-vāiya-dasāo*, (10) *Paṇha-vāgarṇāiṃ*, (11) *Vivāgasuyo*, and (12) *Diṭṭhivāya* (or *Dṛṣṭivāda*) which is not extant or forgotten and thus lost.

The (corresponding) twelve *Upāṅgas* are [2]:

Ovavāiya; *Rāyapaseṇaiya*; *Jīvābhigama*; *Pannavaṇā*; *Sūriya-paṇṇatti*; *Jaṃbuddīva°*; *Candra°*; *Nirayāvaliyāo*; *Kappavaḍaṃsiyāo*; *Pupphiyāo*; *Puppa-cūliyāo*; and *Vaṇhidasāo*.

Then there are ten *Prakīrṇaka*s, six *Mūla-sūtra*s, six *Cheda-sūtra*s and two *Cūlikā*s namely *Nandī-sūtra* and *Anuyoga-dvāra-sūtra*. But there is difference of opinions regarding the number of *āgamas* included in the various categories, e.g., thirty-two *Prakīrṇaka*s (instead of ten) are said to be listed in *Jaina-granthāvali* (1907 CE) [3].

Also, we had 14 *Pūrva*s which were containing the whole of the sacred knowledge (*śrutajñāna*) but which are now lost [4]. However, around the first century CE portions of *Pūrva*s were retrieved by Puṣpadant and Bhūtabali to compose the *Ṣaṭkhaṇḍāgama* and also by Guṇadhara to compose the *Kasāya-pāhuḍa* [5].

After the time of lord Mahāvīra (about 500 BCE), the traditional *śruta-jñāna* (sacred knowledge) and his teachings were being carried on orally by his pupils (*gaṇadharas*). In order to preserve the originality (e.g. *mūlapāṭha*) of the *Āgamas* and continue this task more affectively, the saint-scholars organised Councils from time to time. Five Councils were held [6]:

(i) At Pāṭalīputra about 300 BCE, presided by Sthūlabhadra.
(ii) At Kumarī Parvata, Orissa (2nd cent. BCE) (details not known).
(iii) At Mathurā (300 – 313 CE) presided by Skandilācārya.
(iv) At Valabhī, simultaneously with (iii), presided by Nāgārjuna.
(v) At Valabhī, Gujarat (454/455 CE), presided by Devardhi Gaṇī.

During the last Council, Jaina *Āgamas* were edited and compiled, and were given some sort of book-forms (even variant readings were noted). Thus, the presently available *Āgamas* are said to be dated 454 CE or so. But obviously their contents are to be regarded much older as supported by other evidences in many cases. For instance, *Prajñāpanā* (= the 5th *Upāṅga*, *Pannavaṇā*) is attributed to Ārya Śyāma [7] who is dated about 150 BCE (and died in 92 BCE). Similarly, *Anuyoga-dvāra-sūtra* is [8] attributed to Ārya-Rakṣita who is roughly placed about 100 CE.

Of course, chronology is always a problem especially when we deal with ancient authors and works. In case of Jainology and ancient Jaina works the problem of claims by different sections are also not easy to sort out. For example, *Tattvārthādigama Sūtra* and its *Bhāṣya* are taken to be authored by two different persons who are placed centuries apart [9], while a detailed study [10] shows that the *Sūtra* and *Bhāṣya* both are by Umāsvāti whose period was about 200 to 300 CE. Kundakundācarya who wrote *Pravacana-sāra* etc., also belonged to the same period. I do not claim that this paper is free from controversies regarding works and their authors and times. I have tried to have a rather balanced view in my presentation.

Śūnya (*suṇṇa*) is the most common word used in India for zero. But in ancient works it was also used in some other senses such as worthlessness, absence (*abhāva*) emptiness (void or nullity) etc. Here some instances are given. Cāṇakya [11] was the secretary (or minister) of the emperor Chandragupta Maurya (c. 300 BCE) who is claimed to be a Jaina. Popularly attributed to Cāṇakya is the *Cāṇakya-nīti* which has the saying *avidyaṃ jīvanaṃ śūnyam* ("Life is worthless without *jñāna* or learning").

In the *Ṭhāṇaṃ* (*Sthānāṅga-sūtra*) V. 21 occurs [12] the word *śūnyāgāra* (*śūṇṇāgāra*) which has been translated as *śūnyā-gṛha* meaning a vacant (or empty) house. The same word is also found used in the *śūnya* Naya (which is one of the 47 Jaina Nayas) as stated in the *Pravacanasāra* in the following words:

Śūnyanayena śūnyāgāravat kevalodbhāsi ("By *śūnya* Naya appears *ekāki* like an empty house") [13].

Pāṇini had used the word *lopa* for his grammatical or linguistic zero and his relevant basic *sūtra* is [14] *adarśanaṃ lopaḥ* (Aṣṭādhyāyī, I.1.60).

That the Pāṇinian '*lopa*' stood for a sort of zero is also indicated by the fact that Pūjyapāda (c. 450 CE) in his *Jainendra Vyākaraṇa* replaced *lopa* by *khaṃ* (which was used for zero) in his relevant sutra [15].

In Jaina 'śukla-dhyāna' (white or pure meditaion), there is a 'śūnya-dhyāna' (zero or non-distinctive meditation). Through this the highest goal of life, *mokṣa* (or salvation), is said to be achieved. In this context a few lines from the work *Jñānasāra* may be quoted [see *Jainendra Siddhānta kośa* (ref. 11 at the end), Vol. IV (1998), p. 32]:

....प्राप्नोति समुच्चयस्थानं तथायोगी स्थूलतः *शून्याम्* (38)

....जिनशासनेभणितं *शून्यं* इदमीदृशं मनुते (41)...

नभः सदृशमपि न *गगनं* तत् *शून्यं* केवलं ज्ञानम् (42)...

इति *शून्य* ध्यानज्ञाने लभते योगी परं स्थानम् (43)... .

Note that *Nabha* and *gagana* mean sky and denote zero as a word-numeral (see below). Synonyms of *ākāśa* (sky or ether) are frequently used as words for *śūnya* (zero) and as word-numerals (*Bhūta-saṃkhyās*). The list is large. The grand vastness of the sky made words like *Ananta* ("endless") to be included in the list, as also *antarikṣa.* One may come across word-pair like *śūnya-ākāśa* (vacuum).

If we rely on the commentator (Hemacandra Sūri) of Jinabhadra Gaṇi's *Viśeṣāvaśyaka-bhāṣya* (c. 600 CE) in which the following words are quoted from Bhadrabāhu II's *Āvaśyaka-niryukti* [16].

Thibugāgāra jahanno vaṭṭo....,

0	+	+	१
+	0	१	+
१	१	0	0

or

0	0	१	१
+	१	0	+
१	+	+	0

(as per or in the manuscript)

Figure 1.1.

o			
o	o		
o	o	o	
o	o	o	ō
o	o	o	o

Figure 1.2.

then the word *Thibugāgāra* may be translated as (something like) 'form or shape of *bindu*'. This will show the use of a small circle as a zero.

The use of the zero symbol (small circle) for some other purpose is also found in ancient Jaina works and manuscripts. Like the use of symbols for 0 and 1 by Piṅgala (in his *Chanda-sāstra*), Vīrasena (c. 800 CE) in his *Dhavalā* commentary [17] on *Ṣaṭkhaṇḍāgama* (see Book IV, 1.5.4) uses symbols 0,1 and + (called *haṃsapada*) to explain a type of *dravya-parivartana* (change in *dravya* or substance) by the diagrams as in Figure 1.1.

Another diagram [18] in *Dhavala*, IV.2.6.9 (Book XI, p. 106) has 14 small circles called *bindu* (or *biṃdū*) (see Figure 1.2). It is stated in the text that the two circles represent *ekendriya-sthiti* and the next three lines denote *tiṇṇi sattabhāga* (3/7 parts) of the *sāgaropama*.

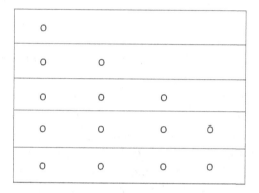

Although *bindu* was used for zero at the time of Vīrasena, here its role is different. The Jaina symbology shown here is perhaps unique.

1.2. All Numbers Expressed in Words (No need of zero)

When no symbols were used, all numbers and fractions were written in words whether they were small or large. In poetry (using verses or *gāthās*) the words were the only choice even if symbols were known for writing the numerals (or digits).

For instance, the traveling of a very large distance in the Aruṇodaya sea is described in the *Bhagavatī Sūtra* II.8.1 (Vol. 1, p. 234) in prose as follows [19]:

'.... *chakaḍiesaepaṇapannaṃ ca koḍio, paṇatīsaṃ ca satasahassāiṃ paṇṇasaṃ ca sahassāiṃ aruṇodaya samudde* ... '

"...six-hundred crore plus fifty-five crore, thirty-five hundred-thousand plus fifty thousand (*yojanas*) in the Aruṇodaya sea..." (which implies the number 655,35,50,000).

As an example of poetry, we quote the following *gāthā* which is found quoted in several ancient Jaina works including the *Jambūdvīpa-prajñapti-sūtra*, VII, 159 (p. 319) [20]

एगं च सय-सहस्सं, तेत्तीसं खलु भवे सहस्साइं ।

णव य सया पण्णासा, तारागण कोडिकोडीणं ।।

"The (number of) *tārāgaṇa* (stars) is certainly one hundred-thousand, thirty-three thousands and nine hundred fifty crore-crore". This means $1,33,950 \times 10^{14}$ stars!

The point to note is that the above wordly statements of numbers do not support any claim of zero or place-value system of numerals.

In studying Jaina religion and philosophy, we come across numbers which are quite big. Very large distances and very long periods are involved in Jaina cosmology and cosmography. Powers of two and squaring is invariably involved. The Jambūdvīpa ('Jambū Island') is surrounded by a series of concentric rings (or annuli) of sea and land alternately and the width of the Nth ring (whether sea or land) is given by

$$W_N = 2^N \cdot D$$

where D (= one lakh *yojanas*) is the diameter of the centrally situated Jambu Island; (by any interpretation the ancient *yojana* measure is far greater than a mile).

Among the non-decimal scales for counting, Jainas were the first to represent large numbers by the successive squares of two, namely

$$1^{\text{st}} \text{ square of } 2 = 2^2$$

$$2^{\text{nd}} \text{ square of } 2 = (1^{\text{st}} \text{ square of } 2)^2 = \left(2^2\right)^2 = 2^{2^2}$$

$$3^{\text{rd}} \text{ square of } 2 = [(2^2)^2]^2 = 2^{2^3} \text{ and so on.}$$

$$\text{Thus the } n^{\text{th}} \text{ square of } 2 = 2^{2^n}, \; n = 1, 2, 3, \ldots$$

It must be noted that in early ancient Jaina works, the above successive squares were all written in words (i.e., without using any symbols). For instance, the *Ṣaṭkhaṇḍāgama* (Book III, Sutra 1.2.45) [21] says that the number of developable human souls lies between *chaṇhaṃ vaggāṇaṃ* (sixth square) and *sattaṇhaṃ vaggāṇaṃ* (seventh square) (of two, in the sense as above). That is, the number is (in modern notation) between 2^{64} and 2^{128}. However, Vīrasena (c. 800 CE) in his commentary (*Dhavalā*, Book III, p. 253) on the above adds that according to other teachers the said number is *verūvassa paṃcamavaggeṇa chaṭṭhamavaggaṃ guṇide* i.e., (it is) product of two's 5th and 6th squares; that is, $2^{32} \times 2^{64}$ (in modern symbols). No concept of zero is involved in these algorithms and even symbology which is used in writing various powers of 2 above is the modern notation (and not found in ancient texts).

1.3. The Decimal and Non-decimal Scales of Counting

Prior to the use of modern numerals (with zero), the decimal scale numbers were written in words in ancient Jaina works. The Prakrit scale was: (1) *ega* (one), (2) *daha* (ten), (3) *saya* (hundred), (4) *sahassa* (thousand), (5) *daha-* or *dasa-sahassa* (ten-thousand),

(6) *saya-sahassa* or *lakkha* (hundred-thousand), (7) *dasa-lakha* (million), (8) *koḍi* (Sanskrit *koṭi*) (ten million) = 100,000,00 (= 10^7), (9) *dasa-* or *daha-koḍi* (ten-crore or 10^8), (10) *saya-koḍi* (hundred-crore), ..., (15) *koḍā-koḍī* (crore-crore, that is crore times crore), and so on, by prefixing other names with *koḍi* which itself was repeated three, four, ... times.

One thing may be pointed out here in this relevant context. Kāccāyana's Pali Grammar [22] has a *koṭi*-scale list of words (for higher counting) from *koṭi* (= 10^7), *pakoṭi* (= 10^{14}), etc. etc., to the last denomination *asaṁkheya* (= $10^{20\times7}$ or 10^{140}). Due to some confusion this high *koṭi* scale series is claimed to belong to *Dhavalā* apparently in *Āsthā aur Cintana* (Delhi, 1987) [23], in Section *Jaina Prācya Vidyās*, p. 62; while the given reference is to A. N. Singh's article (translated into Hindi) included in the Prastāvanā part (only) to the work (Book V). The list is Buddhist and found elsewhere also (see *Gaṇita-tilaka* Baroda, 1937, Introduction, p. XLIX) [24].

However, the Jainas have similar high lists of non-decimal scales in terms of time periods. The 1st name in the series is *pūrvāṅga* which is of 84 lakh years = 84,00,000 years. Successive multiple of this will give the other names. The list from *Anuyoga-dvāra-sūtra* (Beawar ed. 1987, p. 289) is [25]: (1) *Pūrvāṅga* (2) *Pūrva* (3) *Truṭitāṅga* (4) *Truṭita* (5) *Aḍaḍāṅga* (6) *Aḍaḍa* (7) *Avavāṅga* (8) *Avava* (9) *Hūhu-aṅga* (10) *Hūhu* (11) *Utpalāṅga* (12) *Utpala* (13) *Padmāṅga* (14) *Padma* (15) *Nalināṅga* (16) *Nalina* (17) *Acchanikurāṅga* (18) *Accha-nikura* (19) *Ayutāṅga* (20) *Ayuta* (21) *Nayutāṅga* (22) *Nayuta* (23) *Prayutāṅga* (24) *Prayuta* (25) *Cūlikāṅga* (26) *Cūlikā* (27) *Śīrṣa-prahelikāṅga* and lastly (28) *Śīrṣa-prahelikā* which is = $(84{,}00{,}000)^{28}$ years!

Similar lists are found in other ancient Jaina works [26]. The names like *ayuta, prayuta, padma* ('lotus') were also used as decuple terms.

In India ten has been the basis for counting since very early days. Decuple terms or names of numbers of the type 10^n (*daśaguṇottara*) were used. Common among such denominational names were *śata* (hundred), *sahasra* (thousand), *lakṣa* (= 10^5), and *koṭi* (= 10^7). In early times the Jainas frequently used repetitions for higher powers e.g., *koṭi-koṭi* for 10^{14}. In *Ṣaṭkhaṇḍāgama* [27] (Book III, sutra 1.2.45, p. 253) the *koṭi* is repeated three and four times to denote 10^{21} and 10^{28} respectively.

When the full decimal place-value system (with zero) came into use, these decuple terms or decimal denominational names served the purpose of calling the corresponding places by the same names e.g., 8th place will be the *koṭi* place. As higher numbers are frequently met in Jaina cosmography, higher lists of decuple terms (or place-names) were designed. These lists also served the practical and commercial purpose. Repetition of names could be avoided.

Śrīdhara's list of 18 decuple terms became quite popular and it became a standard list. It is found in his *Pāṭīgaṇita* (verses 7-8) as well as in his *Triśatikā* (verses 2-3) which was more popular among Jainas. These names of *daśa-guṇottara* series are as follows [28]:

eka, daśa, śata, sahasra, ayuta (= 10^4),
lakṣa (or *lakṣya*), *prayuta, koṭi, arbuda*(= 10^8),
abja (or *abda*), *kharva, nikharva, mahāsaroja* (= 10^{12}),
śaṅku (or *śaṅkha*), *saritāṁ-pati, antya, madhya* (= 10^{16}),
and *parārdha* (= 10^{17}).

These very eighteen names (except *mahāsaroja* for which we have *mahāmbuja*) are found in the *Abhidhāna-citāmaṇi* (III, verses 537–538) according to H. R. Kapadia [29]. This Sanskrit Dictionary was composed by Hemacandra in 12th century CE [30].

About a century after Śrīdhara (c.750 CE) came Mahāvīrācārya who was the greatest mathematician of his time. In his famous *Gaṇita-sāra-saṅgraha* (I. 63-68), he gave a list of 24 place-names of decimal denominations starting with *eka* which he calls *Prathama-sthāna* (1st place). His full list of 24 *daśaguṇottara* place-names is as follows [31]:

(1) *eka*, (2) *daśa*, (3) *śata*, (4) *sahasra*, (5) *daśa-sahasra*, (6) *lakṣa*, (7) *daśa-lakṣa*, (8) *koṭi*, (9) *daśa-koṭi*, (10) *śata-koṭi* ($= 10^9$); (11) *arbuda* ($= 10^{10}$), (12) *nyarbuda*, (13) *kharva* ($= 10^{12}$), (14) *mahā-kharva*, (15) *padma* ($= 10^{14}$), (16) *mahā-padma*, (17) *kṣoṇī*, (18) *mahā-kṣoṇī* ($= 10^{17}$); (19) *śaṅkha* ($= 10^{18}$), (20) *mahā-śaṅkha*, (21) *kṣityā* ($= 10^{20}$), (22) *mahā-kṣityā*, (23) *kṣobha*, and (24) *mahā-kṣobha* ($= 10^{23}$).

Now 24 is a sacred number for Jainas; so Mahāvīrācārya thought it right to stop at the name of the 24th decimal place-value after giving a sufficiently large table. Also, it should be noted that although the list is long, he could carry out the task by employing fewer words (than one would expect) by resorting to the prefixes *daśa* and *mahā* ('big' or 'great').

It should be noted that Mahāvīrācārya's list is not an extension of Śrīdhara's list because the terms of the same or similar names (beyond *koṭi*) have different values or denominations, and this is regrettable.

Anyway, enthusiasm of Jaina scholars continued in this direction. Rājāditya (12th century CE) in his *Vyavahāra-gaṇitam* (in Kannada) [32] extended Mahāvīrācārya's list to 40 places, the last name being *Mahā-parimita* ($= 10^{39}$). The *Amalasiddhi* gives another list of 97 terms which go(es) upto the name *daśa-ananta* ($= 10^{96}$)! [33].

1.4. The Place-value Method of Counting and Recording, and the Case of Anuyoga-dvāra-sūtra

The *Anuyoga-dvāra-sūtra* (= *ADS*) is a *Cūlikā-sūtra* in Jaina *āgamas*. It is attributed to Āryarakṣita who lived in the 1st century CE Here the purpose is not to discuss the date and authorship of *ADS*, but to present a crucial statement from the work regarding the possible implications in connection with the place-value system of counting. I will also explain the ancient system of decimal counting.

Actually, the *ADS* talks of a particular large number (say N) which occupies 29 (decimal) places, and it is also expressed in terms of powers of two. So, we start with a relevant discussion about that number found elsewhere.

In the *Ṣaṭkhaṇḍāgama* (I. 2.45) (Book III, p. 253) the number of human beings is given as follows [34]:

कोडाकोडाकोडीए उवरि कोडाकोडाकोडाकोडीए हेट्ठदो छण्हं वग्गाणमुवरि सत्तण्हं वग्गाणं हेट्ठदो ।

koḍākoḍākoḍīe uvari koḍākoḍākoḍākoḍīe heṭṭhado chaṇhaṃ vaggāṇamuvari sattaṇhaṁ vaggāṇaṁ heṭṭhado.

"(The number) is above crore times crore times crore and below crore × crore × crore × crore; it is bigger than the sixth *varga* and less than the seventh *varga* (of two)"

Here *varga* means successive square of two, the nth square being 2^{2^n}. Thus, the number of human beings is stated to be more than 2^{64} and less than 2^{128}. It is also estimated to be between

$$(100,000,00)^3 \text{ and } (100,000,00)^4.$$

The correct number of human beings meant here is known to be 2^{96} from other and subsequent sources (see below) [35]. The upper limit in the decimal estimation, namely 10^{28}, is slightly wrong. The correct and closer limits (and even correct value) can or could be found by the decimal place-value of counting and recording (see below). Next, we take the case of *ADS* [36].

In reply to a question about human population, the *Anuyogadvāra-sūtra*, *Sūtra* 423 (Beawar, 1987, p. 349) states [37]:

... जहण्णपदे संखेज्जा संखेज्जाओ कोडीओ, एगुणतीसं ठाणाइं, तिजमलपयस्स उवरिं चउजमलपयस्स हेट्ठा, अहवणंछट्ठो वग्गो पंचमवग्गपहुप्पण्णो, अहवणंछण्णउतिछेयणगदाइरासी,... ।

... Jahaṇṇapade saṁkhejjā saṁkhejjāo koḍīo, eguṇatīsaṁ ṭhāṇāiṁ, tijamalapayassa uvariṁ caujamalapayassa heṭṭhā, ahavaṇaṁ chaṭṭho vaggo paṁcama-vagga-pahuppaṇṇo, ahavaṇaṁ chaṇṇa-uticheyaṇagadāirāsī ...

"In *jaghanyapada*, the number (of human beings) is numerable and is in several crores; it occupies twenty-nine places (in decimal counting). That number is above three *yamala-padas* and below four *yamala-padas*. It is the product of the sixth-square and fifth-square, or it can be halved (or divided by two) ninety-six times".

Now if we square 2^2, the result is called second square of 2 and if we square the result again it is called 3rd square of two. It can be seen that, in this way the nth Jaina square of two is 2^{2^n}.

According to an interpretation [37], *yamalapada* means second successive power of two (i.e., taking $n = 2$, it will be 2^4). For *bi-yamalapada*, we have to take $n = 4$; for *tri-yamalapada* we have to take $n = 6$; for *tetra*; $n = 8$ etc. Thus, the above *Anuyogadvāra-sūtra* (*Sūtra* 423) says that the number of human beings is between 2^{64} and 2^{256}.

Then, the text also says that the number is equal to

$$2^{64} \times 2^{32} (= 2^{96}) = N, \text{ say.}$$

Of course, this can be divided by two as many as 96 times! The text clearly says that N occupies 29 (decimal) places (*ṭhāṇāim*) not digits (*aṅkas*). The text's statement is correct but still it does not necessarily imply the knowledge the modern decimal place-value system of numerals. Ancient way of getting number of places of N is dealt below.

The ancient decimal place-value method of counting and recording is briefly described now. In order to count things or operations, definite places are marked for units, tens, hundreds etc. Suppose we have to count certain operations which are going on one after another. Let the four circles P, Q, R, and S mark or denote the places for units, tens, hundreds, and thousands.

Use of sufficient number of small pebbles will be made in counting the ongoing operations (which are the things to be counted here).

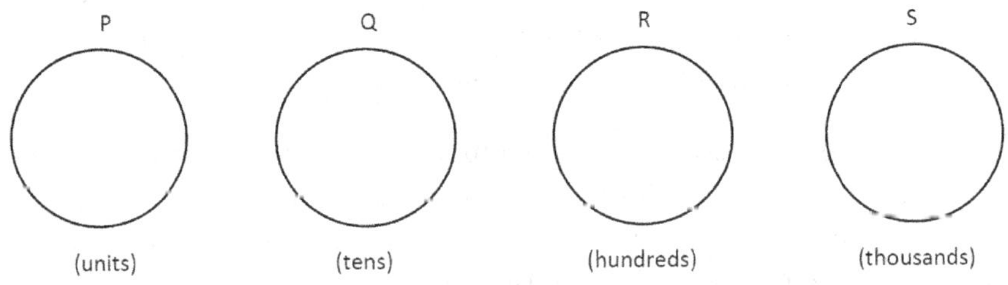

Figure 1.3.

When one operation is over (or counted), we place one pebble (•) in the units circle P. When the second operation is over, we place a pebble again in the circle which will now have two pebbles (to show the count of two operations so far). We repeat the process till there are ten pebbles in the circle P. At this stage we take out all the ten pebbles from P but at the same time place one pebble in the next circle Q (and this will show counting up to ten).

When the next operation is over (or counted), we place a pebble in circle P. The process will be repeated till there are ten pebbles in P. Again, at this stage, we take out the ten pebbles of P but in lieu of these we place a pebble in the circle Q (which will now have two pebbles to show counting up to twenty now). We repeat this counting process till there are ten pebbles in Q. At this stage we take out the ten pebbles from Q but (at the same time) place one pebble in the hundreds circle R. When the next operation is over (i.e., counted) one pebble will be placed in P. In this way we can continue the process of counting the going-on operations.

The above method of decimal place-value counting must be clear now. Each time when the counting of one operation takes place, one pebble is put in circle P. When ten pebbles get collected in any circle, we take them out and (in lieu of them) place just one pebble in the next circle (whose denominational value is ten times higher). The counting can be continued till all operations to be counted (or things to be counted) are exhausted.

The result of counting up to any stage (in the middle of counting or at the end) can always be known by the position of the pebbles at that stage (i.e., by number of pebbles in the various circles). For example, let the position be as shown below in Figure 1.4.

Figure 1.4 shows the number two thousand, three hundred and eight (units). The Q circle is vacant (i.e., absence of a pebble there shows 'no tens' at the stage). Note that for illustration of the method, we just took here *left to right* system; that is, units, tens, hundreds, etc., places begin (or start) from *left*.

The significant point to note is that the method is called place-value because the pebble takes the value according to the place

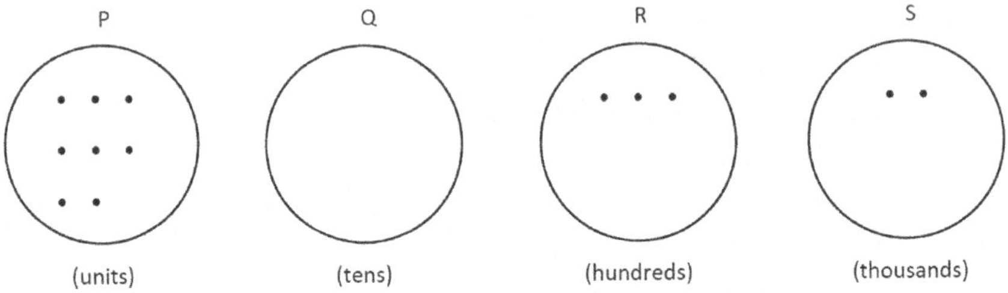

Figure 1.4.

(circle) in which it is placed. This was so clearly stated by Vasumitra (1st century CE) about 2000 years ago as [38]:

"When the clay counting-piece (*vartikā* or pebble) is placed in the place of Units, it is denominated 'one', (when placed in place of Tens, it is denominated as 'ten'), when placed in the place of Hundreds, it is denominated 'hundred', and in place of Thousands, it is denominated a 'thousand'. Similarly............ which makes it indicative of varying numbers."

If the result (of the above method) of place-value counting is to be noted or carried out (say by writing) on paper, it will be convenient to use strokes or small lines (*rekhā*) (instead of pebbles) as markers or counters. In that case the result of counting as in Figure 1.4 will appear (*after shifting to usual* ***right to left*** convention) as in Figure 1.5

Support for this is also there in ancient Sanskrit passage [39]

यथैका रेखा शतस्थाने शतं दशस्थाने दशैकं चैकस्थाने ।

yathaikā rekhā śatasthāne śataṁ daśasthāne daśaikaṁ caikasthāne.

"Just as the same stroke (*rekhā*) is termed a hundred in the hundreds place, ten in the tens place, and one in the units place."

To say that the use or occurrence of figures 1.4 or 1.5 or sayings of Vasumitra or the above passage clearly imply the knowledge of modern decimal place-value system of numerals (with zero) is not correct. A few more steps were needed in this direction. The most important step was to use the numerals (in any form) for numbers

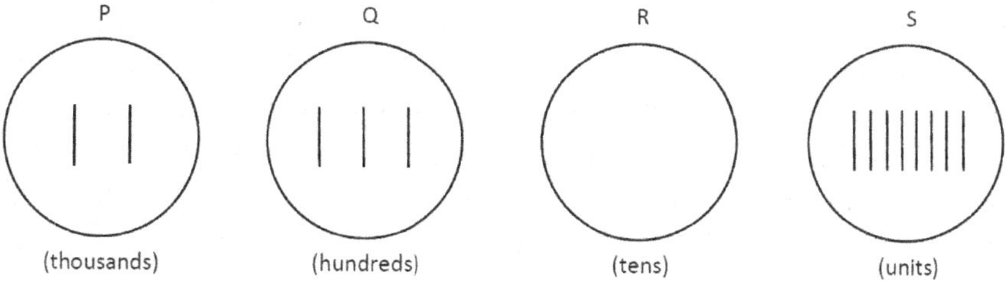

Figure 1.5.

Figure 1.6.

one to nine, say 1,2,3,4,5,6,7,8, and 9. We might use small plates or tokens marked with numbers to be used for *additively gathered* pebbles or strokes. That is, use [1] for |, [2] for ||, [3] for ||| and so on up to [9], as soon as *each* counting is done. In this way, we shall get and record the result as in Figure 1.5 each time and get the output as in Figure 1.6.

Obviously, a symbol is needed to represent absence of a pebble or stroke in a circle or in more circles which mark the places of counting. Thus, 0 was used and we may think of putting a plate [0] in the empty circle in Figure 1.6. With this, we have achieved a method to represent the result of counting at any stage such that *each circle will have only one number or zero symbol* (0). (Note that this was not possible with Figure 1.4 or (Figure 1.5). So, if we strictly follow the order (units, tens, hundreds, etc.) and leave no gaps (for any possible doubts), we get the new system shown by [2][3][0][8] OR simply 2308.

Now we have seen above that the author of the *ADS* knows that the big number N occupies 29 decimal places. That is, it can also be said that N's number of places lies between 3 groups of eight-places and 4 groups of eight-places (as 29 lies between 24 and 32). The text also knows that N is (in modern notation) 2^{96}. But he does not mention the usual 29 digits or digital-presentation of N. It seems that the full decimal place-value system (with a zero symbol) was yet to be known, at least to the author of *ADS*. The number 29 of decimal places could be found by the ancient decimal place-value counting method detailed above. In this method we do not need any symbol for zero.

On the other hand, zero is an essential part of the modern place-value system of numeral (to any base). And it plays its three roles successfully. The three roles are (i) medial or internal, (ii) terminal, and (iii) initial.

1.5. Triloka-Prajñapti and Some Other Early Works

The *Tiloya-paṇṇattī* (or *Triloka-prajñapti*) of Yativṛṣabha is an important early Jaina text. According Hiralal Jain [40], the work and its author may be placed between 473 CE and 609 CE. But it certainly contains material which mostly belong to earlier centuries. Actually, it seems to be a sort of composite work in which portions belong to later times also. It makes full use of the decimal place-value system of numerals developed and already in use (in say about 200 CE).

A significant work *Loyavibhāga* was composed by Muni Sarvanandī in the middle of the 5th century CE. It is not extant in original (but is quoted in *TP*). However, its Sanskrit version *Lokavibhāga* by Siṁhasūri is available [41].

The *Tiloya-paṇṇattī* (= *TP*) (or *Triloka-prajñapti* in Sanskrit) attributed to Yativṛṣabha is a comprehensive work, quite rich in content regarding the Jaina cosmography. Its chapter IV deals with Human World and is titled *Manuṣyaloka-prajñapti.* Portions relevant to mathematics and number-systems are extracted below from the chapter IV [42]:

अइवट्टो मणुव-जगो, जोयण-पणदाल-लक्ख-विक्खंभो ।।६।।

Aivaṭṭo maṇuva-jago, joyaṇa-paṇadāla-lakkha-vikkhaṁbho(6b)

"The Human World is circular and of diameter fortyfive lakh *yojana*" (IV.6b). It is written as *jo 45 la.*

जगमज्झादो उवरिं, तब्बहलं जोयणाणि इगिलक्खं ।

णव-चदु-दुग-ख-त्तिय-दुग-चउरेक्केक्कं-कमेण तप्परिही ॥७॥

Jagamajjhādo uvariṁ, tabbahalaṁ joyaṇāṇi igilakkhaṁ|

Ṇava-cadu-duga-kha-ttiya-duga-caurekkekkaṁ-kameṇa tapparihī ||7||

"Its height above the central part of the universe (world) is one lac *yojana.* The digits in its circumference are (to be read from right to left) nine, four, two, zero, three two, four, and one respectively." (IV.7) It is written in the text as *1 la14230249.*

That is, the Human World has (in *yojanas*)

$$\begin{aligned} \text{Diameter, } D &= 4,500,000 \\ \text{Height, } H &= 100,000 \\ \text{and circumference } C &= 14,230,249. \end{aligned} \tag{5.1}$$

The rule for finding C is given in *gātha* 9 (see below). For the area of the Human World, we have the text:

सुण्ण-णभ-गयण-पण-दुग—एक्क-ख-तिय-सुण्ण-णव-णह्य-सुण्णं ।

छक्केक्क जोयणां चिय अंक-कमे मणुवलोय खेत्तफलं ॥८॥

।१६००९०३०१२५०००।

Suṇṇa-ṇabha-gayaṇa-paṇa-duga-ekka-kha-tiya-suṇṇa-ṇava-ṇahā-suṇṇaṁ|

Chakkekka joyaṇāṁ ciya aṅka-kame maṇuvaloya khettaphalaṁ ||8||

"(Reading right to left) zero, zero, zero, five, two, one, zero, three, zero, nine, zero, zero, six, and one are the digits (*aṅkas*) respectively in (square) *yojana* measure of the area of the Human World. (In figures) 16,009,030,125,000."

That is, the area of the Human World is

$$A = 16,009,030,125,000 \text{ (square) } yojanas. \tag{5.2}$$

TP (IV.9) now gives the traditional ancient Jaina rules for finding the circumference and area of a circle:

वासाकदी दसगुणिदा, करणी परिही च मंडले खेत्ते ।

विक्खंभ-चउब्भाग-प्पहदा सा होदि खेत्तफलं ।।९।।

Vāsākadī dasaguṇidā, karaṇī parihī ca maṇḍale khette |

Vikkhaṃbha-caubbhāga-ppahadā sā hodi khetta-phalaṁ ||9||

"The square-root of the diameter-square multiplied by ten is the circumference of a circular field. That (circumference) multiplied by one-fourth of the diameter becomes the area of the field"

That is,

$$\text{Circumference, } C = \sqrt{10D^2} \tag{5.3}$$

$$\text{and Area, } A = C.(D/4) \tag{5.4}$$

Thus, in the case of the Human World ($D = 45$ lacs) we have

$$\begin{aligned} C &= \sqrt{10 \times (4500000)^2} \\ &= \sqrt{(14230249)^2 + 13397999} \\ &= 14230249 \end{aligned} \tag{5.5}$$

in whole integers (i.e., neglecting fractional part). This is same as given in the text (IV.7). See (5.1).

Using above value of C, the correct rule (5.4) will yield

$$\begin{aligned} A &= (14230249) \times \left(\frac{4500000}{4}\right) \\ &= 16,009,030,125,000 \end{aligned}$$

which is also same as given in the text (IV.8) above.

The next *gātha* (IV.10) of this section on Human World gives its volume as

अट्ठट्ठाणे सुण्णं, पंच-दु-इगि-गयण-ति-णह-णव-सुण्णा ।

अंवर-छक्केक्कं, अंक-कमे तस्स विन्दफल ||१०||

|१६००९०३०१२५००००००००|

Aṭṭhaṭṭhāṇe suṇṇaṁ panca-du-igi-gayaṇa-ti-ṇaha-ṇava-suṇṇā |

Anvar-chakkekkaṁ, aṅka-kame tassa vindaphala||10||

The *gātha* begins by saying that "the first eight places have zero (*suṇṇaṁ*)" and then the remaining eleven digits are given by the usual word-numeral (*bhūta-saṃkhyā*) system from right to left. The formulas (5.3) and (5.4) are ancient Jaina rules found frequently, e.g. in Umāsvāti's *Bhāṣya* (3rd century CE, see p. 2 above) III.11. However, another formula for the area A found in *TP*, IV.2761 shows that we have quoted more ancient part of the work. More accurate calculations in subsequent part also confirm this.

1.6. After Tiloya-paṇṇattī and Upto About 850 CE

The name of Śrīdhara or Śrīdharācārya (8th century CE) was quite popular in ancient Indian mathematics. He is claimed to be Hindu by some and to be a Jaina by other modern scholars. It is also suggested (as a compromise) that he may be accepted as a Hindu who was later on converted to Jainism. Anyway, our concern here is to deal with his mathematical contributions which are relevant to our topics of discussion.

Śrīdhara's following two works are well known [43].

(i) *Pāṭīgaṇita* (or *Bṛhat-pāṭīgaṇita*) which is said to have about 900 verses. Unluckily, it is only partly extant.
(ii) *Triśatikā* (or *Gaṇita-sāra*) which is said (by the author himself) to be extracted or based on (i).

That Śrīdhara was recognized as a great mathematician is clear from the following verse found in the manuscripts and some commentaries of *Triśatikā* [44].

उत्तरतो सुरनिलयं दक्षिणतो मलयपर्वतं यावत् ।

प्रागपरोदधिमध्ये नो गणकः श्रीधरादन्यः ।।

"From the abode of Gods (Himalaya) in the north to the Malaya mountain in the south and between the eastern and western seas, there is no mathematician other than Śrīdhara".

Jinabhadra Gaṇi is a famous scholar of the early 7th century CE. The decimal-place-value system was affectively used by him in describing large numbers in words. Such numbers had large number of terminating zeros.

Vīrasena (early 9th century CE) wrote his voluminous commentary *Dhavalā* on the *Ṣaṭkhaṇḍāgama* sutras. He also started writing the *Jaya-dhavalā* commentary which is on the *Kasāyapahuḍa* of Guṇadhara and which was completed by Jinasena in the first half of the ninth century CE.

Śrīdhara has given some rules regarding the involvement of zero in the four fundamental arithmetical operations. In this *Pātīgaṇita* (= *PG*), Rule 21, he states

क्षेपसमं खं योगे, राशिरविकृतः खयोजनापगमे ।

खस्य गुणनादिकेखं, सङ्गुणने खेन च खमेव ||२१||

Kṣepasamaṁ khaṁ yoge, rāśiravikṛtaḥ khayojanāpagame |

Khasya guṇanādikekhaṁ, saṅguṇane khen ca khameva ||21||

"When something is added to zero, the sum is equal to the additive (*kṣepa*). When zero is added to or subtracted from a number, the number remains unchanged. In multiplication and other operations on zero, the result is zero. Multiplication (of a number) by zero also yields zero."

Thus, the following rules are found here:

(i) $0 + x = x$
(ii) $x \pm 0 = x$
(iii) $0 \times x = 0$

and

(iv) $0/x = 0$, etc.,
(v) $x \times 0 = 0$

The significant thing to note that $(0 + x)$ and $(x + 0)$ and similarly $(0 \times x)$ and $(x \times 0)$ are considered separately. Commutative laws and their awareness are reflected in this.

Śrīdhara rightly avoided to attempt about $x/0$. However, the ancient commentator says [45]

शून्येन यो राशिः सङ्गुण्यते *वा विभज्यते* स शून्यरूप एव

"A number multiplied by or ***divided*** by zero is zero"

That is, $x \times 0 = 0$ ***or*** $x/0 = 0$!

The above *PG*, Rule 21, is also reproduced word by word by Śrīdhara in his *Triśatikā* as Rule 8. But the contents of *PG* as listed by the author himself (see *PG* verses 2-6, p. 2) show that his section on *Śūnya-tattva* [Mathematics & Mathematical philosophy of zero is lost (i.e., not extant)].

The Jaina mathematician Mahāvīrācārya wrote his *Gaṇita-sāra-saṅgraha* (= *GSS*) in Sanskrit in the middle of the ninth century CE. His patron was king Amoghavarṣa (815-877 CE) who was ruling at Mānyakheṭa (South India) and it was a peaceful period in general. The *GSS* is a big work on mathematics (more than 1100 verses) and has nine chapters [46].

The *GSS* is called '*saṅgraha*' or "collection" and is therefore a significant work and source of ancient Jaina mathematics. However, Mahāvīrācārya does not specify the names of the ancient authors and works which he consulted. Also, in spite of the title of *GSS*, the work contains many original contributions of the author. It was used as a text-book for centuries.

Mahāvīrācārya knows the fully developed decimal place-value system with zero. He describes the names of 24 decuple terms, or the names of the decimal denominational places, starting with *eka* (or unity) to *mahā-kṣobha* (*GSS*, I. 63–68). He liberally uses the word-numerals (*bhūta-saṁkhyās*) system and gives a list of such word-numerals for numbers (digits) 1 to 9 and also for zero (*GSS*, I. 53-62). His list of names of zero includes (*GSS*, I. 62): *ākāśa, gagana, śūnya, ambara, kha, nabha, viyat, ananta, antarikṣa, viṣṇu-pāda*, and *div*.

Some rules regarding arithmetical operations involving zero are given in *GSS*, I.49 as follows:

ताडित्तः खेन राशिः खं, सोऽविकारी हृतो युतः ।

हीनोऽपि खवधादिः खं, योगे खं योज्यरूपकम् ।।

Tāḍitaḥ khena rāśiḥ khaṁ, so'vikārī hṛto yutaḥ |

Hīno'pi khavadhādiḥ khaṁ, yoge khaṁ yojya-rūpakaṁ ||

"A number multiplied by zero is zero, and that (number) remains unchanged when it is divided by, combined with (or) diminished by zero. Multiplication and other operations in relation to zero (give rise to) zero; and in the operation of addition, the zero becomes the same as what is added to."

That is, the following rules are implied:

(i) $x \times 0 = 0$
(ii) $x/0 = x$ (see below)
(iii) $x + 0 = x$
(iv) $x - 0 = x$
(v) $0 \times x = 0$, and etc. (e.g., $0 \div x = 0$)
(vi) $0 + x = x$

The Indian *parikarmāṣṭaka* (eight algorithms) includes squaring, cubing, square-rooting and cube-rooting. So, the words 'other operations' (see no. (v) above) in the rule may be taken to include

$$0^2 = 0,\ 0^3 = 0,\ \sqrt{0} = 0,\ \text{and}\ \sqrt[3]{0} = 0.$$

An important point to note is the distinction made between $(x+0)$ and $(0+x)$, and also between $(x \times 0)$ and $(0 \times x)$. The commutative property is not presumably taken to be implied automatically. Zero is the meeting point of the positive and negative numbers. Regarding a negative number Mahāvīrācārya says (*GSS*, I.52b)

ऋणं स्वरूपतोऽवर्गो यतस्तस्मान्नतत्पदम् ॥

Ṛṇaṁ svarūpato'vargo, yatastasmānnatatpadam ||

"By its own nature, a negative (quantity or number) is non-square (*avarga*), therefore it has no square-root".

That is, the square-root of a real negative number is not real or does not exist among real numbers. The remark is quite significant in the history and evolution of imaginary (or complex numbers).

Mahāvīrācārya's rule (ii) is not correct. But why and how he says that a number remains unchanged when divided by zero. I try to answer. Division is a sort of distribution. Let there be N things to be distributed equally among D persons. Then N/D will give things received by each person. On the other hand, if D is the number of things to be given to each person then N/D will be the number of persons eligible or capable to receive their share. But if D is zero, none of the above two type of distribution or division will take place or can be accomplished. That is, when no person is there ($D = 0$) or nothing is to be given ($D = 0$), things N will remain as they are i.e., N will not change. So, $N/0 = N$!

Also, it may be pointed out that translation of the line

'ताडितः खेन शशिः खं, सोऽविकारी हृतो'

is given differently by R. N. Mukherjee [47] and is as follows: "A number multiplied by zero is *zero* and this (*zero* so obtained) remains unchanged when it is divided by zero" (I have slightly changed translation). That is,

$$x \times 0 = 0 \,(\text{which is same as (i)})$$

$$\text{and } 0/0 = 0 \,(\text{this is a new interpretation}).$$

However, the last result is also not correct as $0/0$ is indeterminate and the (limiting) value may vary. Nevertheless, the lapse is minor for a work written in ancient time and it will ever remain a great work of Jaina Mathematics and for History of Mathematics in general.

References and Notes

[1] Jagadish Chandra Jain, *Prākṛta Sāhitya kā itihāsa* (History of Prakrit Literature, in Hindi), Chowkhamba, Varanasi, 1961; p. 34.

[2] *Ibid.*, p. 34.

[3] *Ibid.*, pp. 34–35, foot-note where details are given.

[4] *Ibid.*, p. 35 where the names of the 14 *Pūrvas* are given in the foot-note.

[5] R. S. Shah, "Lure (Lore) of Large Numbers", *Arhat Vacana*, Vol.21, No. 4 (2009), pp. 61–78; p. 62. Actually, Ācārya Dharasena was the source of śṛuta-jñāna for the authors of *ṣaṭ-khaṇḍāgama* (=Chakkhaṃdāgamo in Prakrit).

[6] *Ibid.*, p. 62, and Anupam Jain, *Ardha-māgadhī Sāhitya meṃ Gaṇita* (in Hindi), Ladnun 2008, pp. 21–23 (Also, the correct name of the place is Valabhī).

[7] See J. C. Jain (ref. 1 above), p. 112.

[8] Anupam Jain, *Jaina Darśana evaṃ Gaṇita* (Jaina Philosophy and Mathematics, in Hindi), Meerut, 2019, p. 37.

[9] *Ibid.*, pp. 37–38.

[10] Sagarmal Jain, *Tattvārtha-sūtra aur usakī Paramparā*, (in Hindi), Varanasi, 1994; Introductory page 8-9.

[11] Jinendra Varni, *Jainendra Siddhānta Koś a* (= JSK), 4 volumes, Bharatiya Jnanapith, Delhi; vol. I (1998), pp. 310, 478, and 480.

[12] Muni Nathmal (editor), *Ṭhāṇaṃ* (with Hindi transl.), Ladnun, 1976, p. 550.

[13] See JinendraVarni (ref. 11 above), Vol. II (1997), p. 523.

[14] R. C. Gupta, "Technology of using Śūnya in India", Pages 19-24 in *The Concept of śūnya* (edited by A. K. Bag and S. R. Sarma), Delhi, 2003, p. 21.

[15] See *Aastha aur Chintana* (आस्था और चिंतन) (Acharyaratna Shri Desh Bhushan Maharaj Abhinandan Granth), Delhi, 1987, *Jaina Prācya Vidyāyeṁ* Section, p. 146.

[16] See H. R. Kapadia (editor), *Gaṇita-tilaka of Śrīpati* (with the commentary of Siṃhatilaka Sūri), Baroda, 1937, Introduction, p. XXI.

[17] Hiralal Jain (editor), ṣaṭ-khaṅḍāgama (with *Dhavalā* commentary), Book IV (or volume IV), Amroti, 1942; p. 330.

[18] Phoolchandra Siddhanta-shastri (editor) *Ṣaṭkhanḍāgama* (with the commentary *Dhavalā*), Book XI, Solapur, 1992, p. 106.

[19] *Bhagavatī-sūtra* or *Vyākhyā-prajñapti-sūtra* ed. By Amarmani and Shri Chanda Surana, Part I Beawar, 1991, p. 234. The Āgama text is *Aṅga* no. 5.

[20] *Jambūdvīpa-prajñapti-sūtra* (*Upāṅga* no.6) edited by Chhaganlal Sastri, Beawar, 1994, p. 319. Muni Kanhaiya Lal (editor), *Gaṇitānuyoga*, Ahmedabad, 1986, p. 433 gives the *gāthā* from *Sūrya-Prajñapti* with details.
[21] P. Siddhanta Shastri (see ref. 8 above), Book III, Solapur, 1993, p. 253.
[22] B. Datta and A. N. Singh, *History of Hindu Mathematics*, Single Volume Edition, Bombay, 1962; Part I, pp. 11–12.
[23] See *Aastha aur Chintana* (ref. 15 above) p. 62 of the same Section.
[24] *Gaṇita-tilaka* (see ref. 16 above), p. XLIX.
[25] *Anuyoga-dvāra-sūtra* (by Ārya-rakṣita) edited by Dev Kumar Jain, Beawar, 1987, p. 289.
[26] See Jinendra Varni's JSK (ref. 11 above), vol. II (1997), pp. 216–217. Also, see the recent booklet Who Invented the Zero? (in Hindi) by Anupam Jain, Meerut 2020, pp. 8-9.
[27] See Shastri (editor) (ref. 8 above), Book III, Solapur, 1993, p. 253.
[28] See (a) *Pāṭīgaṇita of Śrīdharācārya* (with an ancient Sanskrit commentary) edited with English translation by K. S. Shukla, Lucknow 1959; p. 5 of text; and (b) *Triśatikā* (or *Pāṭīgaṇita-sāra*) with English translation by V. D. Heroor, Ernakulam, 2015, p. 1.
[29] See Kapadia (ref. 16 above) p. XLVIII.
[30] *Arhat Vacana* Vol.12, No.3 (2000), p. 22.
[31] The *Gaṇita-sāra-saṅgraha* (with Kannada translation and English translation (of M. Rangacārya), edited by Padmavathamma, Hombuja (Karnataka), 2000, pp. 18–19.
[32] *Vyavahāra-gaṇita* (in Kannada) edited by M. M. Bhat, Madras, 1955, p. 3. Also see R. C. Gupta, "World's Longest Lists of Decuple Terms", *Gaṇita Bhāratī* 23 (2001), 83–90.
[33] See *Gaṇitānanda* (Selected works of R. C. Gupta) edited by K. Ramasubramanian, Springer Nature, Singapore, 2019, pp. 113–115.
[34] See the *Ṣaṭkhaṇḍāgama* (with *Dhavalā*), edited by Phoolchandra Siddhanta Shastri, Book III, Solapur, 1993, p. 253.
[35] *Ibid.*, p. 253 where the *Dhavalā* commentary (by Vīrasena of 9th cent.) clearly says that the number is product of 5th and 6th successive squares of two i.e., product of 2^{32} and 2^{64} (see section 2 of the article).
[36] *Anuyoga-dvāra-sūtra* (ref. 25 above), *Sūtra* 423, p. 349.
[37] R. S. Shah, "Mathematics of *Anuyoga-dvāra-sūtra*", *Gaṇita Bhāratī*, 29 (2007), 81-100; pp. 83–84. Also, the term *yamala-pada* may be taken to represent a group of 8 places (not of 8 digits) for convenience of counting of large number of places.

[38] Vasumitra's words are quoted by Śāntarakṣita's pupil Kamalaśīla in his commentary on his guru's *Tattva-saṅgraha*. See English translation (of *Tattva-saṅgraha*) by G. Jha, Vol.II (Baroda, 1939) p. 862. Also, see Hindi translation of Datta and Singh (ref. 22 above) by K. S. Shukla, Lucknow, 1974, p. 72.

[39] See Hindi translations of similar passages in K. S. Shukla (ref. 38), p. 73, along with their reference to original works.

[40] See H. L. Jain and A. N. Upadhye (editors), *Tiloya-Paṇṇattī*, Two vols., Solapur Vol.I (1956), Vol.II (1957), Prastāvana in Vol.II, p. 15.

[41] *Loka-Vibhāga* edited by Balachandra Siddhanta-Shastri, Sholapur, 1962.

[42] *Tiloya-paṇṇattī*, translated into Hindi by Visuddha-mati, Vol.II, 1986.

[43] For editions with English translations of these two works see ref. 28 above.

[44] See *Pāṭīgaṇita* ed., by K. S. Shukla (ref. 28), Introd. p. xxvi

[45] See *Pāṭīgaṇita* (of Śrīdhara) (ref. 28 above), text, p. 14.

[46] For edition and English translation of *GSS*, see ref. 31 above.

[47] R. N. Mukherjee, *Discovery of Zero and Its Impact on Indian Mathematics*, Calcutta, 1991; p. 197.

https://doi.org/10.1142/9789811255502_0002

Article 2

Evolution and Development of Mathematical Ideas through Jaina Scriptures

Ratnakumar S. Shah

301 Matruchhaya, 40 Jeevanchhaya Society
Paud Road, Pune 411038, Maharashtra, India
ratnakumarshah@yahoo.co.in

The literature (scriptures) of Jainism, while primarily concerned with rituals and metaphysics, involves many mathematical ideas. Their cogitation of the nature of the Universe led to the development of geometric concepts, recurrence relations, arithmetic and geometric and more advanced series. Their concept of karma as super fine particles interacting with the soul led to the introduction of rules of indices, permutations and combinations, binomial coefficients, probability, logarithms, and theory of transfinite numbers. Their atomic theory enabled them to have a brush with partition function of modern number theory. They were able to deal with some advanced topics such as occupancy problem of probability theory, algebraic symbolism etc. In the canonical work *Anuyogadvārasūtra* [2], we find evidence suggesting that the concept of zero having place value and the first glimpse of transfinite numbers somewhat similar to the Cantorian treatment of transfinite numbers (although lacking in rigor).

This paper is based on the Canon pertaining to period from 4th C. BCE to 5th C. CE and commentaries thereon belonging to the period from 6th C. to 15th C. CE.

Keywords: Decimal place value system, zero, square roots, types of infinities, permutations and combinations, mensuration of the circle, algebraic identities, logarithms.

Mathematics Subject Classification 2020: 01-06, 01A32, 01A99.

2.1 Decimal System with Place Value.

2.1.1 The Decimal system based on 10 was prevalent in India since antiquity. As early as in *Yajurveda Saṃhitā* (1200 – 900 BCE), the decimal scale goes from *eka* (1) to *daśa* (10), *śata* (10^2), *sahasra* (10^3), *ayuta* (10^4), *niyuta* (10^5), *prayuta* (10^6), *arbuda* (10^7), *nyarbuda* (10^8), *samudra* (10^9), *madhya* (10^{10}), *anta* (10^{11}) and *parārdha* (10^{12}). The epic *Rāmāyana* also mentions very large numbers of the order of 10^{62} while describing the army of its hero *Rāma*. [5] (p. 242). Buddhist mathematicians devised names for powers up to 10^7 (*koṭi*) and then for powers of *koṭi* up to 20, i.e 10^{140} (*asaṅkhyeya*). Probably, they included all numbers $\geq 10^{140}$ in the class of innumerable (*asaṅkhyāta*). [6] (Book 4; p. xviii, article by Prof. A. N. Singh).

2.1.2. Decimal Scale in Jaina Scriptures.

2.1.2.(a). In the Jaina scriptures, the scale of numeration used predominantly is the decimal scale with place value. *Anuyogadvārasūtra* [2] (sutra 139), of 5th C. gives the scale as *eka* (one = 1), *dasa* (ten = 10), *sata* (hundred = 10^2), *sahassa* (thousand = 10^3), *dasasahassa* (ten thousand = 10^4), *sayasahassa* (hundred thousand = 10^5), *dasasayasahassa* (million = 10^6), *koḍī* (ten million = 10^7), *dasakoḍī* (hundred million = 10^8), *sayakoḍī* (billion = 10^9) and *dasasayakoḍī* = 10^{10}). Further, sutra 190 mentions that a *sāgaropama* (sea-measure) is equal to 15 *koḍākoḍī* times a *palyopama* (pit-measure). As we have seen, *koḍī* = 10^7, then *koḍākoḍī* = $10^7 \times 10^7 = 10^{14}$ and thus, the scale goes up to hundred trillion. In *Ṣaṭkhaṇḍāgama* [21], about 150 years earlier, there is a mention of *koḍākoḍākoḍākoḍī* which extends the scale to 10^{28}.

2.1.2.(b). Large Numbers.

The Jaina scholars had a great penchant for very large numbers. *Tiloyapaṇṇatti* (1/2 C.) gives a mixed-decimal temporal scale going up in decimal scale up to 100000, and 84 times of that is called *puvvaṅga* (=8400000 years). 8400000 times of *puvvaṅga* is *puvva* (8400000 × 8400000 = 70560000000000 years). Then multiplying alternately by 84 and 8400000 the scale goes up to *acalātma* or *acalappa* $= (84)^{31} \times 10^{90}$ and is described as: "put 84 in 31 places, multiply them together and add ninety zeroes to arrive at the *acalappa*". [23] (vol. 1, ch. 4, g. 308, p. 178). This is a number with 150 digits. [$\approx 4.49 \times 10^{149}$]. In [23], however, are furnished different names for all the 28 places in the above process. Ages of very important personalities in the hoary past are given in these units. E.g., the life-span of the first *Tirthaṅkara*, *Ṛṣabhadeva*, is stated to be 8400000 *puvva* (= 592704000000000000000 years!]. [23] (vol. 1, g. 579, p. 214).

In [2] (sū.137) and many other *Aṅgas* and *Upāṅgas* (c.4th C.), there are 28 stages from *puvvaṅga* to *sīsapaheliyā*, each 8400000 times of the previous one so that the *sīsapaheliyā* $= (8400000)^{28} =$ $758263253073010241157973569975696406218966848080183296 \times 10^{140}$ (a 194-digit number). In fact, the full expansion with 194 digits was first time given by *Ā. Abhayadeva* (11th C) in his commentary on *Ṭhāṇāṅga*. However, as we will see below, *Jinabhadra* (5/6th C) had that prowess to perform these calculations. For, *Jinabhadra* gives the number of stars in the family of one moon as: "*chhāvaṭṭhi sahassāiṁ, nava ceva sayāiṁ pañcasayarāiṁ ... koḍākoḍīṇaṁ*" (i.e., 66975×10^{14}), [3] (vol. 2, p. 296), and since in the Half-*Puṣkara* Island there are 72 Moons, the number of stars there is given as 4822200×10^{14}. [3] (vol. 2, p. 296).

According to *Ạcharya Akalaṅka* (7th C.) of *Digambara* tradition, *acalātma*, the highest enumerable temporal unit $= (8400000)^{31}$. [14] (vol. 2, p. 216, 217). The value of *sīsapaheliyā* of [2] given above is what was revised in Valabhi Conference of c.450 CE In fact, [2] was revised in Conference held simultaneously at Mathura and Valabhi in c.300 CE As per the Valabhi version then *sīsapaheliyā* $= (8400000)^{36}$ $\approx 1.87 \times 10^{249}$.

2.2. The Concept of Zero with Place Value

2.2.1. The discovery of zero and the symbol for it is shrouded in mystery. Right from the oldest scripture, *Ācārāṅgasutta*, the word for zero, *suṇṇa* (Pr.) or *śunya* (Sk.), meant 'empty'. In ch.10, sū. 490 of [2] we find a very interesting statement about the minimum number of corporeal (*audārika*) bodies of human beings. It states that the minimum number of physical bodies of the human beings satisfies the following 4 criteria:

i) It is a number having 29 places (*ṭhāṇāiṁ*) and is equal to several *koḍākoḍī* ($= 10^7 \times 10^7 = 10^{14}$) (*saṅkhejjao koḍākoḍio*).

ii) This number is bigger than '*tiyamalapaya*' and less than '*cauyamalapaya*'. Now, '*yamalapaya*' stands for a number with 8 digits, and hence *tiyamalapaya* is a number with $3 \times 8 = 24$ digits and *cauyamalapaya* is a number with $4 \times 8 = 32$ digits. In i) above it is already stated that the number is with 29 digits. **iii)** The number can be obtained by the multiplication of 5th successive square of 2 with 6th successive square of 2. Now here, by first square of 2 means simply $2^2 = 4$. Then 2nd sq. of 2 will be $4^2 = 16 = 2^4 = 2^{2^2}$; 3rd sq. of $2 = 16^2 = 256 = 2^8 = 2^{2^3}$; 4th sq. of $2 = 256^2 = 65536 = 2^{16} = 2^{2^4}$; 5th sq. of $2 = 65536^2 = 4294967296 = 2^{32} = 2^{2^5}$; and 6th sq. of $2 = 184466814016 = 2^{64} = 2^{2^6}$. Then, if we denote by H the minimum human population, the statement amounts to saying that $H = 2^{2^5} \times 2^{2^6} = 2^{32}$ x $2^{64} = 2^{96} = 79\ 228\ 162\ 514\ 264\ 337\ 593\ 543\ 950\ 336$. **iv)** This number can be successively halved 96 times. This is obvious from iii) above since $H = 2^{96}$.

2.2.2. From (i) and (iii) above it is clear that when *Anuyogadvārasūtra* (see [2]) was edited in 300 CE the Jaina mathematicians were aware that it is a number with 29 digits and the number at the thousand's place is 0 (zero)

2.2.3. The idea in (iv) ultimately culminated in evolution of the operator called *ardhaccheda* which is equivalent to the modern log_2. Similarly, the concept in (iii) above resulted in evolution of the operator *vargaśalākā* which is same as the modern $log_2 log_2$. Both these concepts were possibly discovered well before 8th C. as *Vīrasena* (8/9 C.) and *Nemicandra* (10/11 C.) use the operators profusely.

2.3. *Sarvanandī.*

There is a *Saṁskṛta* work titled *Lokavibhāga* (Sections of the Universe) by *Acharya Simhasūrī Ṛṣi* who in a short colophon at the end of the work has stated: "This work is the mere transliteration in *Saṃskṛta* of the ancient work in *Prākṛta* by *Ācharya Sarvanandī*, who wrote it in

Śaka Era 380 [458 CE]". From his testimony the original *Prākṛta* work must belong to 5th C. CE. [18], (preface, p. 16, 25; colophon, p. 225). There are numerous instances of recounting of big numbers from right to left which clearly include zero at certain place/s. *Sarvanandī* recounts many large numbers in his tome digit-wise from right to left using the words '*kramāt*' (in order), '*sthānakaiḥ*' (place-wise), '*sthānakramāt*' (in place-order), '*aṅkakrmāta*' (digits in the order of), '*vistṛtāḥ*' (when expanded) etc. He gives the diameter of 8th island, *Nandiśvaradvīpa*, as 13107200000 *yojana* and describes in words from right to left as 'five zeros then two, seven, sky, one, three and face' (*pañcabhyaḥ khalu śūnyebhyaḥ paraṁ dvai sapta cāmbaraṁ / ekaṁtrīṇi ca rupaṁ* ...//). It may be noted that, for the reason of poetics, sometimes certain words were used for the digits, e.g., sky for zero and face for one in the present case. And there are large number of such instances in *Lokavibhāga*. [18] (p. 79). This shows that the decimal place value system with zero having place value was firmly entrenched in common use in 5th C. CE and that the numeration and symbols must have been invented at least a century or two earlier.

2.4. *Jinabhadra.*

2.4.1. *Jinabhadragaṇi Kṣamāśramaṇa* (5/6 C. CE) was a great saint-scholar of *Śvetāmbara* tradition, a prolific writer and his works were regarded as almost Agamic in nature. He shows the same proficiency in arithmetical operations as is available in modern days (of course, sans calculators and computers). He tackles large numbers with ease, defines quite large numbers on place-value basis or expresses them digit-wise, evaluates roots of very large numbers, and handles areas and volumes of certain types of figures most imaginatively.

2.4.2 (a). In his work [3], up to *gāthā* 68 of *Adhikāra* (chapter) 1 there is no mention of zero. In g.68 he calculates the square of the northern chord of Mt. *Vaitāḍhya* as 41490097500 *kalās* (1 *kala* = 1/19 *yojana*) and expresses it as '*sattanaui-sahassa pañcasaya aunāpannaṁ koḍīigayālisaṁ koḍisayā* (ninety-seven thousand five hundred and forty-nine *koḍīs* and forty-one thousand *koḍīs*). Here the number is given in words based on decimal place-value scheme only and there is no mention of zero directly.

2.4.2 (b). In the very next *gāthā* [3] (g. 69), he determines the square of the northern chord of the Continent of *Bharata* as 75600000000 *kalā*s and reads it (from left to right) as '*paṇasayaro chacca aṭṭhasunnāiṁ*' (seventy-five six and eight zeros). In the same *gāthā* he calculates the square of the northern chord of Mt. *Cullahimavanta* as 224400000000 *kalā*s and reads it as '*du visa coyala sunnaṭṭhaṁ*' (twenty-two, forty-four, eight zeros). See also [25] (p. 402–419).

2.5. Square-roots and Compound Fractions.

With the discovery of zero, it seems considerable proficiency was obtained in extracting square-roots of very large numbers and tackling quite some compound fractions. As early as in 4th C. they were able to find out the chord of Mt. *Mahāhimavāna* by using the formula for the chord of a circle as $c = \sqrt{4h(d-h)}$, where d is the diameter (of *Jambudvīpa*, our earth regarded as shaped like a circular disc) $= 100000$ *yojana*, h is the (smaller) height of the chord and c the chord (here length of Mt. *Mahāhimavāna*). And the problem reduces to evaluating the radical: $c = \sqrt{4 \times \frac{150000}{19} \times (100000 - \frac{150000}{19})} = 53931\frac{6}{19}$ *yojana*. Although the formula is not stated, the correct value is stated and there are numerous calculations of this type in [13] (ch. 4, sū.62). Same is the case with other scriptures [23], [18], [3] etc, except that the general formula for the chord is clearly stated in [3] (g. 35, 36)

2.6. The Number Field: Z^+.

The number field of Jainas was that of positive integers, the modern Z^+. Negative numbers were not recognized. As early as in *Bhagavatīsūtra* [4], this field was divided in 2 parts: the *jumma* (Sk. *Yugma*) or even numbers and the *oja* or odd numbers. There is another classification in 4 ways of the number field in [4]. The *kaḍajumma* (*kṛtayugma*) numbers of type $4n$ or $\equiv 4 \bmod 4$ [i.e. 4, 8,12, ...], *kallyoja* numbers of type $4n+1$ or $\equiv 1 \bmod 4$, [i.e. 1, 5, 9, ...], *dāvarajumma* (*dvāparaugma*) numbers of type $4n+2$ or $\equiv 2 \bmod 4$ [i.e 2, 6, 10, ...], and *teyoja* (*trayoja*) numbers of type $4n+3$ or $\equiv 3 \bmod 4$ [i.e. 3, 7, 11, ...]. Probably, the concept of zero was not yet fully developed then. Hence, in [4] *kadajumma* numbers are of the type $4n + 4$ or $\equiv 4 \bmod 4$. Anyhow, we may say this as a sort of beginning of modular arithmetic. [4] (Book 4, p. 693).

There is yet another type of classification in [4] based on divisibility by 16 and leaving the remainder as 1, 2, 3, ..., 15, 16 (or zero), i.e., numbers of type $16n+1$, $16n+2$, ..., $16n+15$, $16n+16$ (or $16n$) or $\equiv 0, 1, 2, \ldots, 15 \bmod 16$. Each of the type is given a special name. Thus *kaḍajumma-kaḍajumma* are the numbers $\equiv 16$ (or 0) mod 16, e.g., 16, 32, 48 etc. *Kaḍajumma-teyoja* are the numbers $\equiv$ 3 mod 16, i.e., the numbers 19, 35, 51 etc. [4] (Book 4, p. 679–681). This is not merely a theoretical exercise, but [4] identifies the real quantities belonging to these 16 classes.

2.7. Ordinals and Cardinals.

2.7.1. There is yet another division in 3 classes of the number field: *saṅkhyāta* (countable or enumerable or numerable), *asaṅkhyāta* (innumerable or uncountable or countless), and *ananta*(infinite, unbounded, limitless). (We shall hereafter use the words in English underlined for the corresponding *Saṃskṛta* or *Prākṛta* words). In the early scripture like [15] of (1st C.) and [16] of ($3/4^{th}$ C.) there is only this 3-fold division. In [23] of ($1^{st}/2^{nd}$ C.) and [21] of (2nd C.) 2 more classes are added giving 5 classes, namely *saṅkhyāta, asaṅkhyāta, asaṅkhyātasaṅkhyāta* (innumerably innumerable), *ananta* and *anantānanta* (infinitely infinite). It is in [2] of (4th C.) we find the mathematical treatment of innumerables and infinites which resembles to some extent Cantor's theory of hierarchy of infinites (in cardinals and ordinals) as may be clear from the next sub-section 7.2.

2.7.2. The *Anuyogadvārasūtra* Model

In [2], the number field is divided in 7 classes, 1 of numerables, 3 of innumerables, and 3 of infinites. There schematic representation is somewhat like this.

Numerables range from 2 (minimum), 3, 4, ..., A − 1 (maximum), where A is minimum of innumerables. Note that 1 was not regarded as a number by Jaina mathematicians as it acts as a unit only for measurement. So, minimum numerable is 2 and maximum numerable is A − 1.

***Parittāsaṅkhyāta* (lower order innumerables)** go from A (min.), A+1, A+2, ..., $A^A - 1 = B - 1$ (maximum) say. ADS describes in detail the procedure for determining A. It is a horrendously large number given by $A \sim 10^{10^{10^{\cdot^{\cdot^{10^{44}}}}}}$, where the height of the tower of 10 is of the order of 10^{40}, and we will express it by Knuth's up-arrow notation as $A \sim 10 \uparrow\uparrow 10^{44}$. This number is comparable with the Graham number or Tree (3) number of modern mathematics.

***Yuktāsaṅkhyāta* (middle order innumerables)** proceed as $A^A = B(\text{min.}), B + 1, B + 2, \ldots, B^B - 1 = A^{A^{A+1}} - 1 = C - 1$ (max.) say.

***Asaṅkhyātasaṅkhyāta* (higher order innumerables or innumerably innumerables)** range thus: $C = B^B = A^{A^{A+1}} \sim A \uparrow\uparrow 3$ (min.), $C + 1, C + 2, \ldots, C^C - 1 = D - 1$ (max.) say. Hereafter, we enter the realm of infinites.

***Parittānanta* (lower order infinites)** go as $D = C^C = B^{B^{B+1}} \sim A^{A^{A^{A+1}}}$ (min.) $\approx A \uparrow\uparrow 4, D + 1, D + 2, \ldots, D^D - 1$ (max.) $= E - 1$ say.

***Yuktānanta* (middle order infinites)** range from $E = D^D = C^{C^{C+1}} \sim A^{A^{A^{A^{A+1}}}}$ (min.) $\approx A \uparrow\uparrow 5, E+1, E+2, \ldots, E^E - 1 = F - 1$ (max.) say.

***Anantānanta* (higher order infinites)** proceed as $F = E^E \sim A^{A^{A^{A^{A^{A+1}}}}}$ (min.), F + 1, F + 2, ... and ADS is emphatic that the sequence continues unabated and that there is nothing like a maximum, hinting at the ordinal infinite of first order of Georg Cantor, the founder of set theory. [7] (p. 86–87).

2.7.3. Estimate of the Minimum of Innumerables

In [2] is described a procedure by which an innumerable number of variable cylindrical pits of uniform depth 1000 yojana, the diameter of first pit being 100000 *yojana*. *Nemicandra* (10/11 C.) in [24] (p. 16) calculates under some assumptions the number of rape seeds that can be filled in the first pit is 199711293845131636363636363636363636363636363 $\frac{4}{11} \approx 1.997 \times 10^{44} \sim 10^{44} = N_0$ say. The diameter of second pit, as per procedure laid down, is $\sim 2^{N_0}$ and the seeds contained within it can be shown to contain seeds whose number would be $\sim 2^{2N_0} = N_1$ say. It was beyond his capacity to determine N_1. But as per the procedure outlined the number of seeds in the 3rd pit would be $\sim 2^{N_2} = N_3$ say. There are N_0 such variable pits (*anavasthā kuṇḍas*) and the number of seeds in the last (N_0-th) pit would be $\sim 2^{N_{N_0-1}} = N_{N_0}$ say. We can immediately see the parallels between the sequences $N_0, N_1, N_2, \ldots$ and the sequence of Cantorian sequence of cardinal infinites $\aleph_0, \aleph_1, \aleph_2, \ldots$ etc., although N_0, N_1, $N_2, \ldots$ are not infinites but are unimaginably gargantuan numbers comparable to modern Graham number or Tree (3) number.

2.8. Integer Partitions.

2.8.1. The theory of partitions of integers was first developed by Euler (18th C) and later by Hardy, Littlewood and Ramanujan towards the end of 19th C and beginning of 20th C. The theory has applications in statistical mechanics and nuclear physics. The partition function, $p\ (n)$, is defined as the number of ways an integer can be represented as the sum of other integers (the integer itself is counted and order of the integers is disregarded). E.g.

$1 = 1$, and $p(1) = 1$;
$2 = 1 + 1$, and $p(2) = 2$;
$3 = 2 + 1 = 1 + 1 + 1$, and $p(3) = 3$;
$4 = 3 + 1 = 2 + 2 = 2 + 1 + 1 = 1 + 1 + 1 + 1$, and $p(4) = 5$; etc.

A few further values of this function are: $p\ (6) = 11$, $p\ (7) = 15$, $p\ (10) = 42$, $p\ (15) = 176$, $p\ (30) = 5604$, $p\ (50) = 204226$, $p\ (100) = 190569292$. From this it can be observed that the function $p\ (n)$ is a fast-growing function.

2.8.2. In [4] (Book1, p. 140–151) considers an almost similar function, namely as to the number of ways a molecule (*skandha*) formed by n ultimate particles (*paramāṇus*) can be split into molecules and atoms. Suppose we call this function $m\ (n)$, then it gives the values of the function $m\ (n)$ for $n = 2, 3, 4, \ldots, 10$ as 1, 2, 4, 6, 10, 14, 21, 29 and **40**. From this it can be immediately inferred that this function in [4] is related with a very simple relation with the partition function $p\ (n)$, e.g., $m(n) = p(n) - 1$, except for m (10) which should be 41 (and not 40). Probably, it must have been an error on the part of the copyist of manuscript. While counting partitions, [BS] doesn't take the original number in account. [In fact, [4] lists all these partitions. E.g., for a molecule with 4 atoms, it can be split in the following 4 ways: 1) a molecule (*skandha*) with 3 atoms (*paramāṇus*) and 1 atom, 2) a molecule with 2 atoms and another molecule with 2 atoms, 3) a molecule with 2 atoms, 1 atom and 1 atom and 4) 4 atoms separately].

2.9. Temporal Scale. Analogy Measures (*Upamāmāna*).

2.9.1. We have seen in sec.1.2(b) above that in [4] and [2] the temporal scale goes up to $(8400000)^{31} \approx 758 \times 10^{193}$ years. Then, there is a significant remark that normal arithmetic stops here and counting further can be done using analogy-measures (*upamāmāna*) only. [4], sū.137. The procedure roughly is to fill up a cylindrical pit of diameter 1 yojana and 1 yojana deep by hair- tips of an infant and removing 1 hair tip per 100 years, which is termed as *Palyopama* and 10^{15} times of the *Palyopama* as *Sāgaropama*. This was the position, probably, prior to 4th C., since [4], [13], [22] speak of only these 2 measures everywhere. However, *Nemicandra* (10/11 C.) and *Malayagiri* (12 C.) were able to determine the number of hair-tips in the pit as 413452630308203177749512192 x 10^{18} ($\approx 4.1345 \times 10^{44}$) and 330762104246562542199609 7-536 $\times$ 10^9 ($\approx 3.3 \times 10^{36}$) respectively. [23] (vol. 1, ch. 1, g.123, 124, p. 15; [2] (sū.204, 205, commentary by *Malayagiri*). The *palyopma* is then defined as the period elapsed if each of the hair-tips are removed per 100 years. Thus, the *palyopama* would be $\approx 4.13 \times 10^{46}$ years or $\approx 3.3 \times 10^{38}$ years respectively.

2.9.2. Later Jaina scholars must have found that these values are far below the *acalātma* or *hastaprahelita* or *śīrṣaprahelikā*. It appears, in [2] onwards, they were required to change these quantities by further dividing each hair-tip in innumerable parts thus taking the *palyopama* and *sāgaropama* to '*asaṅkhyāta*' (innumerable) class. They also inserted 3 classes each for *palyopama* and *sāgaropama*, namely, *uddhāra palyopama*, *addhā palyopama* and *kṣetrapalyopama*, and *uddhāra/addhā/kṣetra sāgaropama* respectively, (each *sāgaropama* being 10^{15} times of the corresponding *palyopama*).

2.10. Occupancy Problem in Probability Theory: (r balls in n cells)

In *Bhagavatīsūtra* [4] is considered an interesting problem in modern probability theory and combinatorics called 'occupancy problem'. The problem of finding out the number of ways $r = 1, 2, \ldots, 10, 11, 12$ soul/souls entering $n = 1, 2, \ldots, 6, 7$ hell/hells or $1, 2, \ldots, 10, 11, 12$ heaven/heavens. This is same as the number of ways r balls (souls) can occupy n cells (hells or heavens). The number of ways r balls can be put in n cells is given by the formula: $A_{r,n} = \binom{n+r-1}{r} = \binom{n+r-1}{n-1}$. [11] (p. 38).

This formula gives the values of number of ways $1, 2, \ldots, 7$ souls entering 7 hells as:

$$A_{1,7} = \binom{7}{1} = 7, \quad A_{2,7} = \binom{8}{2} = 28, \quad A_{3,7} = \binom{9}{3} = 84 \quad ,$$

$$A_{4,7} = \binom{10}{4} = 210, \quad \ldots, \quad A_{7,7} = \binom{13}{7} = 1716.$$

In [4], are enumerated and described these combinations manually. [4] (vol. II, p. 461–485). It is not certain whether the formulas for combinations were known then. It is quite probable that by the time of Mathura recension (c.300 CE) the theory of combinations and permutations was fully developed, since all the canonical literature was first revised then. [4] contains many more references which show that combinatorial techniques were applied to explain some philosophical ideas. [4], [2] etc contain the term '*bhaṅgasamukkittana*' which is equivalent to modern 'combinatorics'. Anyhow, in [4] are recited all the combinations as given above.

Commentator *Abhayadeva* (10/11 C) then tries to determine number of ways $r = 1, 2, \ldots, 10, 11$ (*saṅkhyāta*), 12 (*asaṅkhyāta*) souls can enter $n = 1, 2, \ldots, 12$ heavens by noticing the recurrence relation $A_{n,r+1} = \binom{n+r}{r+1} = \binom{n+r-1}{r} \times \frac{n+r}{r+1}$.

Hence, once $A_{12,1} = \binom{12}{1} = 12$ is determined, then $A_{12,2} = \binom{13}{2} = 12 \times \frac{13}{2} = 78$, $A_{12,3} = 78 \times \times \frac{14}{3} = 364$, $A_{12,4} = 364 \times \frac{15}{4} = 1365$, $A_{12,5} = 1365 \times \frac{16}{5} = 4368$, $A_{12,6} = 4368 \times \frac{17}{6} = 12376$, $A_{12,7} = 12376 \times \frac{18}{7} = 31824$, $A_{12,8} = 31824 \times \frac{19}{8} = 75582$, $A_{12,9} = 75582 \times \frac{20}{9} = 167960$, and $A_{12,10} = 167960 \times \frac{21}{10} = 352716$. [4] (vol. II, p. 461–485).

2.11. The Geometry and Mensuration of a Circle and Concentric Annuli.

2.11.1. We come across full-fledged geometry of circle and concentric annuli in [13] of 4th C or earlier and other scriptures as the very subject matter is *Jambuddīva*, our earth, which is described to be a circular flat disc with 7 continents and 6 mountain ranges occupying alternately the east-west running parallel chords. There are calculations of length of chords, arcs, areas etc. in these scriptural works, especially in [18], [3], [23], [24] etc. They contain calculations of extraction of square roots of very large numbers. As the diameter of *Jambuddīva* is 100000 *yojana*, the circumference is given as 316227 *yojana* 3 *kosa* 128 *dhaṇus* and little over 13 ½*aṅgulas*. Dividing this by 100000 *yojana* and converting to *yojana* the smaller units we get a value very close to $\sqrt{10}$ in conformity with other calculations of chords etc. where the Jaina scholar-saints took π to be $\sqrt{10}$. This is known as Jaina value of π and was even used by *Brahmagupta* in his calculations. (*Āryabhaṭa* of 5th C. has given the value of π as $\frac{62832}{20000} = 3.1416$ in [1] (p. 146)).

All the works mentioned above contain the use following formulas in respect of a circle. Suppose, diameter of the circle is d, length of chord is c, length of circumference is C, height of the segment formed by the chord is h, length of the arc of segment is s, then

(i) $C = \sqrt{10}d$,
(ii) area of circle $A = \frac{1}{4}C.d = \frac{1}{4}\sqrt{10}d^2$, [23] (vol. 1, p. 143).
(iii) $(\frac{c}{2})^2 = \sqrt{(\frac{d}{2})^2 - (\frac{d}{2} - h)^2}$, [23] (vol. 1, p. 163).

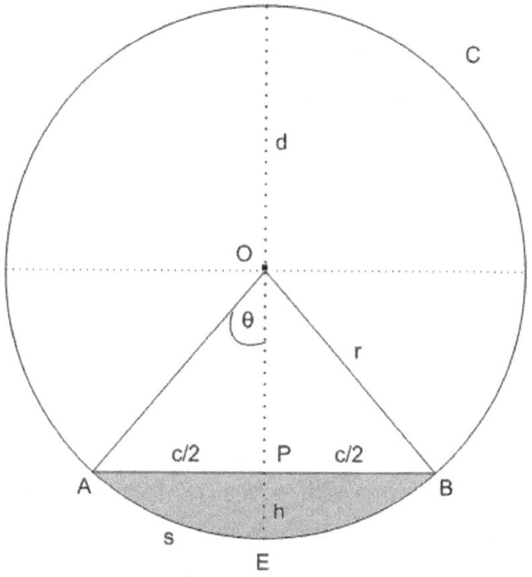

(iv) $c^2 = 4h(d-h)$, [3] (vol. 1, p. 91).
(v) $s = \sqrt{c^2 + 6h^2}$, [3] (vol. 1, p. 95).
(vi) Area of segment, A $= \frac{\sqrt{10}}{4} \times c \times h$.

These formulas were specifically stated in [23] and [24] in word-form and not in symbols. Formula at (v) appears to have been arrived at empirically, but the maximum error in calculations of arcs of various continents never exceeds 1.3%. *Jinabhadra* gives the formula for area A between 2 parallel chords c_1 and c_2, $(c_2 > c_1)$, i.e., areas of 7 continents and 6 dividing mountains, with heights of respective segments h_1 and h_2 as: $A = \sqrt{\frac{c_1^2+c_2^2}{2}} \times (h_2 - h_1)$. [3], (vol. 1, p. 130). This formula (stated in word-form) was probably arrived at empirically gives results with high accuracy. *Jinabhadra* then proceeds to calculate the volumes of these dividing mountains and continents.

2.11.2. Geometry and Mensuration of annuli: Our Earth, *Jambuddīva*, is regarded as a flat circular disc is supposed to be

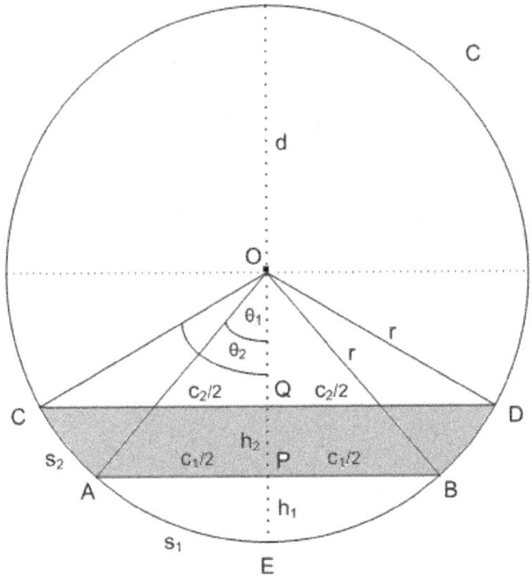

surrounded by innumerable annular seas and islands, with width of each going on doubling. [23] gives 18 formulas relating to their widths (w_n), diameters (d_n) and areas as number of times the area of *Jambuddīva* (N_n).

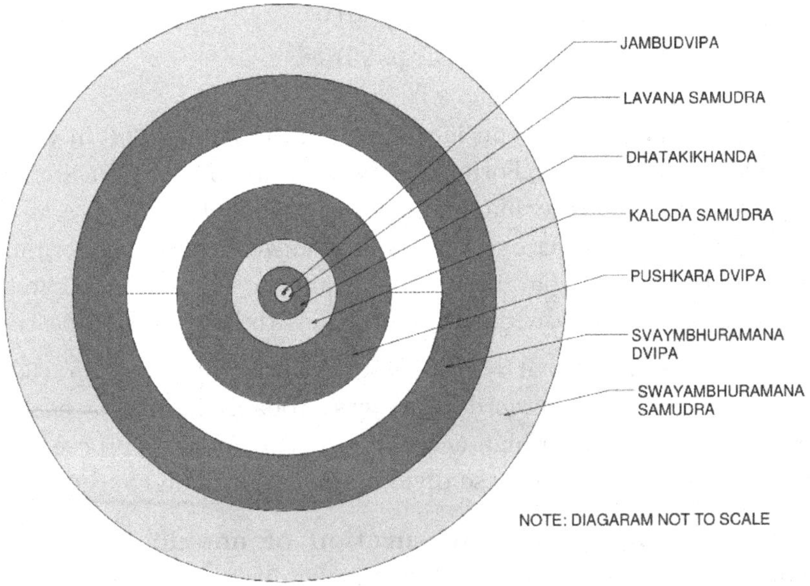

2.11.3. The central island *Jambudvīpa* is 100000 y, the width of next *Lavaṇa* Sea is 200000 y, that of next island *Dhātakīkhaṇḍa* is 400000 y, that of next *Kāloda* Sea is 800000 y and so on. The following 4 formulas can be easily observed and proved.

(A) $w_n = 2w_{n-1}$, (B) $w_n = 2^{n-1}$, (C) $d_n = 1 + 2(2 + 4 + \ldots, +2^{n-1}) = (2^{n+1} - 3)$, (D) $d_n - d_{n-1} = 2w_n$.

There are 18 formulas stated in [23] (vol. 2, g. 245–264, p. 566–579), all in word-form, can be proved easily by using 4 formulas [(A) to (D)] stated above. As an illustration, we give only 4 of them garbed in modern notation.

(a) $w_{n+1} - (w_n + w_{n-1} + w_{n-2} + \cdots 1) = w_{n+1} - \sum_{k=1}^{n} w_k = 1$, (g. 245; p. 566),
(b) $\frac{d_{n-1}}{2} = \frac{w_n - 2 + d_n}{5} - \frac{1}{2}$, (g. 246-247; p. 567),
(c) $\frac{1}{2} + 2 + 4 + \cdots + 2^{n-1} = \sum_{k=1}^{n} w_k = \frac{d_n}{2}$, (g. 249; p. 568),
(d) $N_n = 3(w_n - 1) \times 4w_n$, where N_n is number of times of area of *Jambuddīva* the area of n-th island or sea is, (g. 261; p. 576).

All the above formulas are stated in word form and we have interpreted it in modern algebraic symbols. To shorten them, we have shown them in the units of 100000 *yojan*.

2.12. Beginnings of Algebra

2.12.1. *Vīrasena* explains many concepts by assigning certain numbers to them and refers to the process as '*aṅkasandṛṣṭi*' (numerical representation) and refers to '*arthasaṅdṛṣṭi*' (algebraic representation) at many places in his encyclopaedic tomes [6] and [12]. Rhetorical algebra appears in Jaina scriptures in the form of '*karaṇa sūtras*' (formulas) and many of them are interspersed in *Āgamic* and post-*Āgamic* literature. We find the following formula in combinatorics quoted in [19] of 2nd C. in word-form: $\binom{64}{1} + \binom{64}{2} + \binom{64}{3} + \cdots + \binom{64}{64} = 2^{64} - 1$, by designating it as '*gaṇita gaha*' (mathematical verse or formula). [19] (Book13, p. 248). Of course, it is stated in word-form.

2.12.2. *Virasena*'s Rhetorical Algebra.

Virasena states hundreds of results in words in his tomes [6] and [12] which makes us believe that some sort of algebraic notation may be prevalent then. Here are some examples.

2.12.2.(a). Remaider Theorem. If dividend is N, divisor is a × b, then division is either (A) c and positive remainder R or (B) $c+1$ and negative remainder r, then

(A) $N = (a \times b) \times c + R$, or
(B) $N = (a \times b)\ (c+1) - r$.
This can also be written as
(A) $\frac{N}{a} = b \times c + \frac{R}{a}$, and
(B) $\frac{N}{a} = b \times (c+1) - \frac{r}{a}$. [6] (Book3, p. 47–48).

2.12.2.(b). Ratios. If $\frac{a}{b} = c$, then **(i)** $b + \frac{b(c-d)}{d} = \frac{bc}{d} = b - \frac{b(d-c)}{d}$, **(ii)** $\frac{a}{b+d} = \frac{bc}{b+d} = c - \frac{cd}{b+d}$, and **(iii)** $\frac{a}{b-d} = \frac{bc}{b-d} = c + \frac{cd}{b-d}$. All these results are stated rhetorically. [6] (Book3, p. 48–50).

2.12.2.(c). Algebraic Identities. If c is the total number of souls (living beings) in the Universe, a is the number of liberated souls, b is the number of souls under different (13 in number) stages of spiritual development (*guṇasthānas*), d is the number of deluded (*mithyādṛṣṭi*) souls, i.e., if $a + b = c$, and further if $\frac{c}{b} = l$, $\frac{c}{d} = m$, then:

(i) $\frac{c}{l+m} = \frac{c}{l(\frac{m}{l})} = \frac{\frac{c}{l}}{1+\frac{b}{d}} = \frac{b}{1+\frac{b}{d}}$, and
(ii) $\frac{c}{l-m} = \frac{d}{\frac{d}{b}-1}$. [6] (Book3, p. 45–47).

2.12.2.(d). While discussing the mechanics of bondage of karmas, *Vīrasena* derives some interesting algebraic relations, viz. **(i)** $\frac{A^2}{A+1} = A - 1 + \frac{1}{A+1}$, **(ii)** $\frac{A^3}{A^2+A+1} = A - 1 + \frac{1}{A^2+A+1}$, **(iii)** $\frac{A^4}{A^3+A^2+A+1} = A - 1 + \frac{1}{A^3+A^2+A+1}$. [6], (Book11, p. 215–217).

2.12.2.(e). A Queer Identity. On several occasions *Ṣaṭkhaṇḍāgama* and other Āgamic works refer to a relationship, e.g., $\frac{world\,line}{innumerable}$ = innumerable *koḍi* yojanas = product of successive square roots of space points in the world-line = innumerable times the space points in the square-root of the world-line. The innumerables in each may be different. In modern symbols, putting the world-line (*jagaśreḍhī*) as λ and the innumerables with suitable suffixes of A, we can write: $\frac{\lambda}{A_1} = A_2 \times 10^7 y = \lambda^{\frac{1}{2}} \times \lambda^{\frac{1}{4}} \times \lambda^{\frac{1}{8}} \times \ldots \lambda^{\frac{1}{2^{n-1}}} \times \lambda^{\frac{1}{2^n}} = A_3 \times \sqrt{\lambda}$.

No proof of the identity was given. However, *Vīrasena* presents a brilliant proof on lines below:
$\frac{\lambda}{A_1} = \frac{\lambda^{\frac{1}{2}}}{A_1} \times \sqrt{\lambda} = \frac{\lambda^{\frac{1}{4}} \times \lambda^{\frac{1}{8}} \times \lambda^{\frac{1}{8}}}{A_1} \times \sqrt{\lambda} = \frac{\lambda^{\frac{1}{4}} \times \lambda^{\frac{1}{8}} \times \lambda^{\frac{1}{16}} \times \lambda^{\frac{1}{16}}}{A_1} \times \sqrt{\lambda} = \frac{\lambda^{\frac{1}{4}} \times \lambda^{\frac{1}{8}} \times \lambda^{\frac{1}{16}} \times \ldots \times \lambda^{\frac{1}{2^N}} \times \lambda^{\frac{1}{2^N}}}{A_1} \times \sqrt{\lambda} = A_3 \times \sqrt{\lambda} = A_2 \times 10^7 y$ for obvious reasons. [6] (Book3, p. 142).

2.12.2(f). Summation of A.P. and G.P. For this he states the formulas for summation of n terms (S_n) of an A.P. with first term a(*ādidhana*), last term l (*antimadhana*) and common difference (*caya*) d, and for a G.P. with first term a and common ratio (*guṇakāra*) r in the word-form, which when put in the modern notation, are equivalent to: For an A.P., $S_n = a \times n + d(1 + 2 + \cdots (n-1)) = a \times n + d \times \frac{(n-1) \times n}{2} = \frac{n(a+l)}{2}$. For a G.P., $S_n = \frac{a(r^n-1))}{r-1}$.

2.12.2(g). Number of the Moons. After working out the operations pertaining to A.P. and G.P., *Virasena* derives the formula for the number of Moons in the k-th island or sea (M_k) as: $M_k = 9 \times 2^{2k-2} + 2 \times 2^{k-1} \times (2^{k-1} - 1) = 9 \times 2^{2k-2} + 2 \times 2^{2k-2} - 2^k = 11 \times 2^{2k-2} - 2^k$ for $k \geq 6$. Then, he illustrates this result by calculating $M_6 = 11200$ and $M_7 = 44928$. He then states the formula for total number of Moons from 6th sea to k-th island/sea as: $\sum_{6}^{k} M_k = \frac{11}{3}(2^{2k} - 2^{10}) - (2^{k+1} - 2^6)$. He verifies by showing $\mathrm{M}_6 = 11200$ and $M_6 + M_7 = 56128$. The number of Moons in the 1st island (*Jambūdvīpa*) to 5th *Puṣkaradvīpa* add to $2+4+12+42+1336 = 1396$. Thus, the number of Moons from 1st to k-th island/sea is $\sum_{6}^{k} M_k + 1396$. *Vīrasena* then asserts that the total number of Moons in the Universe (*Loka*) is: $\frac{11}{3}(2^{2N} - 2^{10}) - (2^{N+1} - 2^6) + 1396$, where N is the number of islands and seas in the Universe. [6] (Book 4, p. 154–156). All these general results are stated in word-form, and for sake of convenience, we have shown them in the modern symbolic notation.

2.13. Emergence of Binomial Coefficients. Six-fold Increase/ Decrease.

12.13.1. According to Karma Theory of Jainas, karmas are in the form of fine (infinitesimal in size) particles which bind the soul. Liberation from the cycle of perpetual births and deaths can be achieved by completely shedding the karma particles attached to the soul. Such processes of bondage (fusion) and shedding (fission) are of exponential nature. These increases or decreases are of 6 types and the process is called '*śaṭsthānapatita vṛddhī/hānī*' (six-fold increase/decrease).

12.13.2. This 6-fold increase decrease is described as: i) Infinite(th)-part increase/decrease (*anantabhāga vṛddhi/hānī*), ii) innumerable(th)-part increase/decrease (*asaṅkhyātabhāgavṛddhi/hānī*), iii) numerable(th)-part increase/decrease (*saṅkhyātabhāga vrddhi/ hānī*), iv) numerable-times increase/decrease (*saṅkhyātaguṇavṛddhi/ hānī*), v) innumerable-times increase/decrease (*asaṇkhyātaguṇa vṛddhi/hānī*), and vi) infinite-times increase/decrease (*ananta-guṇa vṛddhi/hānī*).

If the infinite-part increase happens *kāṇḍaka* (= k say) times then innumerable-part increase happens once. Then infinite-part increase will start and after k times again innumerable-part increase will take place. In this manner when innumerable-part increase takes place k times, first time numerable-part increase will take place. Then, process will resume with infinite-part increase followed by innumerable-part increase followed by numerable-part increase, and when numerable-part increase takes place k times, numerable times increase will take place for the first time. Increase process will restart from infinite-part increase, then innumerable-part increase, numerable-part increase, numerable-times increase, so that when numerable-times increase takes place k times, innumerable-times increase will take place for the first time. Similarly, beginning anew from infinite-part increase, then innumerable–part increase, numerable-part increase, numerable times increase, innumerable times increase, when this innumerable times increase takes place k times, it results in infinite- time increase. All this is described in detail in [19] (Book 4; p. 171–173).

12.13.3. Thus, when initial infinite-part increase takes place k times, (i.e., at k+1-th place), innumerable-part increase will take place. Similarly, from initial infinite-part increase, it will take further total $k + k(k) = k + k^2 = k(\mathbf{1} + \mathbf{1}.k)$ steps to arrive at numerable-part increase. Total number of steps after initial infinite-part increase to arrive at numerable times increase is then $k + k^2 + k(k + k^2) = k^3 + 2k^2 + k = k(\mathbf{1}.k^2 + \mathbf{2}.k + \mathbf{1})$, for innumerable- times increase the process will have further steps $= k^3 + k^2 + k + k(k^3 + k^2 + k) = k^4 + 3k^3 + 3k^2 + k = k(\mathbf{1}.k^3 + \mathbf{3}.k^2 + \mathbf{3}.k + \mathbf{1})$ and total further steps to have infinite-times increase from initial infinite-part increase is $k^4 + 3k^3 + 3k^2 + k + k(k^4 + 3k^3 + 3k^2 + k) = k^5 + 4k^4 + 6k^3 + 4k^2 + k = k(\mathbf{1}.k^4 + 4.k^3 + 6.k^2 + 4.k + \mathbf{1})$. All these expressions are described in the word-form.

We can see that in describing exponential process of increase in bondage of karmas, Jaina scholar-saints stumbled across binomial coefficients, as shown in bold letters above, although they did not generalise it further..

2.14. Binomial Coefficients and Power Expansion.

Vīrasena may be aware of the binomial coefficients emerge in the expansion of

$$(1+a)^n = 1 + \binom{n}{1} a + \binom{n}{2} a^2 + \binom{n}{3} a^3 + \cdots + \binom{n}{r} a^r + \cdots + \binom{n}{n-1} a^{n-1} + \binom{n}{n} a^n$$

This assertion is confirmed by the calculations he made in to determine 3^{10} in [12] (Book 4, p. 309) in the following manner:

$3^{10} = (1+2)^{10} = 1 + 10 \times 2 + \frac{10\times 9}{1\times 2} \times 2^2 + \frac{10\times 9\times 8}{1\times 2\times 3} \times 2^3 + \cdots + \frac{10!}{10!} \times 2^{10} = 1 + 10 \times 2 + 45 \times 4 + 120 \times 8 + 210 \times 16 + 252 \times 32 + 210 \times 64 + 120 \times 128 + 45 \times 256 + 10 \times 512 + 1 \times 1024 = 59049$. [12] (Book 4; p. 309). He also immediately illustrates this method to show $3^9 = 19683$, $3^{11} = 177147$ and $3^{13} = 1594322$. [12] (Book 5, p. 177).

2.15. Combinatorics.

2.15.1. The first work to give the general formula to determine the combinations of r things out of n is '*Pañcakalpabhāṣya*' [19a] a commentary on *Kalpa* by *Saṅghadāsagaṇi* (6th C), In g.915, 916 a general formula is stated which is equivalent to:

$$\binom{n}{r} = \frac{n\,(n-1)\,(n-2)\ldots(n-r+1)}{1.2.3.\ \ldots r}.$$

In g.928, the combinations $\binom{16}{r}$ for $r = 1, 2, \ldots, 16$ are worked out as 16, 120, 560, 1820, 4368, 8008, 11440, 12870, and for the remaining combinations it states that the they are the same numbers in reverse order (i.e., 11440, 8008, 4368, 1820, 560, 120 and 16). This means that there was the knowledge of the identity: $\binom{n}{r} = \binom{n}{n-r}$.

Gāthā 920, 921 clearly state in words a formula which amounts to modern formula:

$\sum_{r=1}^{n} \binom{n}{r} = 2^n - 1$. This last formula must be known to Jaina scholars about 4 to 5 centuries earlier as [2] of (2nd C.) quotes an ancient *gāthā* stating that the maximum number of words in the scriptural knowledge formed by combination of 1, 2, 3 etc. up to 64 out of 64 vowels, consonants, colophons etc. is $2^{64} - 1$ (= 18,446,744,073,709,551,615), although the figure in the brackets is given in its commentary [6] (Book 13, p. 248).

2.15.2. In Book 2, (p. 300 and 308) of [12], *Vīrasena* further illustrates this for calculating each of $\binom{n}{r}$ by using the recurrence relation: $\binom{n}{r+1} = \binom{n}{r} \times \frac{n-r}{r+1}$, in the following manner:

$\binom{64}{1} = \frac{64}{1} = 64$, $\binom{64}{2} = 64 \times \frac{63}{2} = 2016$, $\binom{64}{3} = 2016 \times \frac{62}{3} = 41664$, $\binom{64}{4} = 41664 \times \frac{61}{4} = 635376$, $\binom{64}{5} = 635376 \times \frac{60}{5} = 7624512$, $\binom{64}{6} = 762452 \times \frac{59}{6} = 74974368$, etc.

2.16. Logarithms (base 2).

2.16.1. This theme was fully developed by *Vīrasena* (8/9 C.) and *Nemicandra* (10/11 C.). *Vīrasena* considers log_2 (*ardhaccheda*), $log_2 log_2$ (*vargaśalākā*), $log_2 log_2 log_2 log_2$ (*vargaśalākā* of *vargaśalākā*) of the number of deluded souls. He employs the term '*ardhaccheda*' which is equivalent to the modern operator log_2 and '*varga-śalākā*' which is same as modern $log_2 log_2$. The *ardhaccheda* (*ac*) of certain sum Q is then defined as index (power) of 2 to arrive at that sum. I.e., if $Q = 2^k$ then $ac(Q) = log_2(Q) = k$. In the same manner if the quantity Q can be obtained by squaring 2 successively k times, then k is the *vargaśalākā* (*vs*) of Q is k. I.e., if $Q = 2^{2^k}$, then $vs(Q) = log_2 log_2(Q) = k$.

2.16.2. Laws of Logarithm.

In [6], Book3, on p. 1, 24, 55, 60, and 345, *Vīrasena*, while discussing about the number of deluded souls, makes use of various formulas for the operations of logarithms with base 2, but clearly many of them are applicable for any other base. These rules are:

1) $log(m.n) = log\, m + log\, n$ (p. 60),
2) $log\ (m/n) = log\, m - log\, n$ (p. 60),
3) $log\, m^n = n\, log\, m$ (p. 55),
4) $log_2 2 = 1$ (p. 55),
5) $log_2\ 2^m = m$ (p. 55),
6) $log\ \{(m^m)^2\} = 2\ m\ log\ m$ (p. 1),
7) $log_2 log_2\ \{(m^m)^2\} = log_2\ (2m\ log_2 m) = 1 + log_2\ log_2 m$ (p. 1),
8) $log\ (l.m.n) = log\ l + log\ m + log\ n$ (p. 1),
9) $log[(m^m)^{m^m}] = m^m log\, m^m = m^{m+1} log\, m$ (p. 24),
10) $log\ \log[(m^m)^{m^m}] = log\, m + m\, log\, m + log\, log\, m(p.\ 24)$.
11) $\frac{P}{A} = \frac{P}{2^{log_2 A}} = \frac{\sqrt{P}}{2^{log_2 A}} \times \sqrt{P}$ (p. 345).
12) $L^{\frac{log_2 T}{log_2 L}} = T$ (p. 345).
13) $L^{\frac{1}{log_2 L}} = 2$ (p. 345).

We find the results 1) to 10) in the works of the later *Acharya Nemicandra* also [24].

2.16.3. Logarithm of Composite Numbers.

We illustrate how *Vīrasena* calculates the logarithm, base 2, of a composite number by successively halving it: $log_2\left(\frac{256}{13}\right)$ $\frac{128}{13}, \frac{64}{13}, \frac{32}{13}, \frac{16}{13} = 1\frac{3}{13}$, hence $log_2\left(\frac{256}{13}\right) \approx 4\frac{3}{13} = 4.23077\ldots$. The correct value is $4.29956\ldots$. Similarly, he calculates $log_2\left(\frac{65536}{13}\right) \approx 12\frac{3}{13} = 12.23077\ldots$. The correct value is $12.29956\ldots$. Not only this but he illustrates how to calculate the logarithm of a composite number with base 3, by successively dividing the number by 3: i.e., $\frac{256}{39}$, $\frac{256}{117}$, $\frac{256}{351}$ and gets the approximate value of $log_3\left(\frac{256}{13}\right)$. In modern parlance, what *Vīrasena* is trying is to solve the equation: $\frac{256}{3^x} = 13$. Solution of this equation gives $x = 2.71372\ldots$, while his value is $x = 2.59259\ldots$, an error of –4.45%. [6] (Book3, p. 59).

2.16.4. Two (Advanced) Inequalities.

Vīrasena asserts that the number of all living beings in the universe, say J, is $>> F$, where F is cardinal infinite of order 3, and it is greater than the number obtained by raising F to its own power successively 3 times. To be more accurate, he specifies the upper bound of J. For this he introduces the term "*vargita-saṁvargita*" meaning raising a quantity to its own power. Thus, first *vargita-saṁvargita* of $F = F^F = G$, say; second *vargita-saṁvargita* of $F = G^G = H$ say; third *vargita-saṁvargita* of $F = H^H = I$, say. Now, *Vīrasena* asserts that $J < I$. To prove this, he discusses 2 inequalities in his [6] (Book 3, p. 10–26).

(1) $G < vs\ (I) < G^2$, and (2) $\frac{vs\{vs(I)\}}{vs(F)} > D$, where $vs \equiv log_2\ log_2$. (For the proof of these identities, see [2a], p. 99.

2.17. Pre-Calculus.

2.17.1. *Natural logarithm of 2, ln2.*

Jaina scholar-saints had always considered the exponential increase (or decrease) of quantities like bondage of karmas (fusion), destruction of karmas (fission) etc. and they were more concerned about when these quantities go on doubling (or halving). Thus, *Virasena* makes a detailed analysis of limit of $(1+\frac{1}{n})^n$ as n grows larger and larger.

2.17.2. *Vīrasena* was considering the binomial equation: $\left(1 + \frac{1}{n}\right)^k = 2$ to determine when the bondage or of karma doubles. he develops binomial series: $\left(1 + \frac{1}{n}\right)^k = 1 + \binom{k}{1}\left(\frac{1}{n}\right) + \binom{k}{2}\left(\frac{1}{n}\right)^2 + \cdots + \binom{k}{k}\left(\frac{1}{n}\right)^k = \sum_{0}^{k}\binom{k}{r}\left(\frac{1}{n}\right)^r$. If n is sufficiently large, $1/n$ is a small fraction (infinitesimal), and if we denote it by Δ, the above expression can be written as:

$(1+\Delta)^k = 1+\frac{k}{1}.\Delta+\frac{k(k-1)}{1.2}.\Delta^2+\frac{k(k-1)(k-2)}{1.2.3}.\Delta^3+\ldots+\Delta^k$. Not only this, but *Vīrasena* interprets first few terms of this series geometrically. He has special names for each power of Δ, viz. *prakṣepa (Δ), piśula (Δ^2), piśulāpiśula (Δ^3), cūrṇikā (Δ^4), cūrṇā-cūrṇi (Δ^5), bhinna (Δ^6), bhinnābhinna (Δ^7), chinna (Δ^8), chinnāchinna (Δ^9), truṭita (Δ^{10}), truṭitā-truṭita (Δ^{11}), dalita (Δ^{12}), and dalitādalita (Δ^{13})*. All this he describes in detail in [6] (Book 12, p. 158–162).

2.17.3. He declares, by trial and error, $(1+\frac{1}{56})^{41} = 2$. This exactly enabled him to find value of *ln* 2. Remember, $2 = (1+\frac{1}{56})^{41} = (1+\frac{1}{56})^{56.\frac{41}{56}} \approx e^{\frac{41}{56}}$ *and*, *ln* $2 = \frac{41}{56}$ $(= 0.732\underline{142857})$, a recurring decimal number. The actual value of *ln* $2 = 0.693147\ldots$. He even determines *ln* 3 by considering the exponential expansion: $(1+\frac{1}{56})^{64\frac{1}{2}} = 3$. It appears that in these cases *Vīrasena* must have considered first 4 terms of the expansion, $\sum_0^3 \binom{64.5}{r}(\frac{1}{56})^r = 3.04774...$ In fact, $(1+\frac{1}{56})^{64\frac{1}{2}} = 3.13$ (*Vīrasena*'s value is only less by 4.3%). And if we write $3 = (1+\frac{1}{56})^{64\frac{1}{2}} = (1+\frac{1}{56})^{56.\frac{64\frac{1}{2}}{56}} \approx e^{\frac{64.5}{56}} \approx e^{ln 3}$. So, *ln* $3 \approx \frac{64.5}{56} = 1.13$, actual value being 1.10 (*Vīrasna*'s value exceeds only by 2.7%). He proceeds to calculate *ln* 4 by asserting: $(1+\frac{1}{56})^{79} = 4$, (actually it is $= 4.048\ldots$). This time, because of higher power, *Vīrasena* would have considered sum of first 5 terms of the binomial expansion, i.e., $\sum_0^4 \binom{79}{r}(\frac{1}{56})^r = \frac{5614443}{1404928} = 3.996...$ The error in *Vīrasena*'s value is just 1.2%! Thus, as per *Vīrasena*'s interpretation, *ln* $4 = \frac{79}{56} = 1.41..$ (actually, *ln* $4 = 1.38629\ldots$, error of + 1.73%). These calculations are explained in word-form in [6] (Book 14; p. 178–185).

2.18. Algebraic Symbolism (*Artha-sandṛṣṭi*).

2.18.1. Syncopated algebra: Use of alphabets or numerals to denote certain quantities or mathematical operations began with commentators of [23], [9], [10] from 11th C onwards. E.g., प or 1 for *paliovama*, सा or 2 for *sāgarovama*, सू or 3 for *sūcyaṅgula* (linear finger), प्र or 4 for *pratarāṅgula* (planar finger), घ or 5 for *ghanāṅgula* (cubic finger), ज or 6 or — for *jagaśreṇi* (world-line) which is equal to 7 *rājus* and written as 7 र , लोप्र or 7 or = for *lokapayara* (world-plane), लो or 8 or ≡ for *loka* (universe), यो for *yojana* etc. With this they could write the equation: र = $\genfrac{}{}{0pt}{}{-}{7} = \genfrac{}{}{0pt}{}{=}{49} = \genfrac{}{}{0pt}{}{\equiv}{343}$, which is equivalent in modern notation to: $R = \frac{L}{7} = \frac{P}{49} = \frac{U}{343}$, where R= *rāju*, $L = 7R$ is measure of length of world-line, $P = 49R^2$ is measure of area of world-plane, and $U = 343R^3$ is measure of volume of the universe. Also note that in ancient notation fractions were written without the separating horizontal line. $1\frac{1}{2}R$ less 100000 *yojana* was shown symbolically as $\genfrac{}{}{0pt}{}{3-}{14}|$ रियो 100000, where रि is the symbol for 'minus', and the volume of nether world $\frac{4}{7} \times 343R^3$ is represented by $\genfrac{}{}{0pt}{}{\equiv}{7}4$. [23] (vol. 1, p. 11, 16, 19–21). After 11th C. we find symbols such as; केमू१ for $\sqrt{\sqrt{kevalajn\bar{a}na}}$, केमू२ for $\sqrt{\sqrt{kevalajn\bar{a}na}}$ etc., $\frac{2}{3}$ means 3+2, $\frac{2-}{3}$ means 3-2, 2×3 is represented as 2 | 3 or 2 गु 3, and $3 \div 2$ as $\genfrac{}{}{0pt}{}{3}{2}$ or 3 भा 2. Further, we find even shorter symbols, e.g., *ardhaccheda* of *Palya*, i.e., $\log_2$ P is denoted by छे, $\log_2\log_2$ P as व. Hence, the relation $log_2 a = log_2\left(P^{log_2 P}\right) = (log_2 P)^2$ is shown as छेछे . [23] (vol. 1, ch. 1, g. 93, 149, p. 11,18 and g. 166, p. 21).

2.18.2. Symbolic Algebra: Commentators *Mādhavacandra Traividya*, *Keśavavarṇī* of GSJ, GSK, KS have further developed the algebraic symbolism while discussing karma theory in these tomes. We give only one example here. They use a symbol resembling modern σ for *sankhyāta* (enumerable number), ∂ for *asankhyāta* (innumerable), व्य is equivalent to sign for minus and स to plus sign. 2σ is used to represent an *antaramuhutta* (a time interval between 2 *samayas* and a *muhutta* less 1 *samaya*). $\equiv \partial$ represents innumerable times the space-points in the universe (U).

18.3. The symbols used in 11-13th C. were somewhat like ones shown in the figure below. Note that multiplication is denoted by writing the multiplicad and multiplicator besides one another as in $\equiv \partial$ or by putting a vertical line between them as $\sigma\,|\,2$.

≡∂ |2 σσσ |2   व्य ≡∂ |2   स ≡∂| १ / (2σσσ| १^ / σ|2)
2σσσ|2σσσ|σ|2   2σσσ|2σσσ|σ|2   2σσσ|2σσσ|σ|2

=≡∂ |2 σσσ |2   स ≡∂| १ / (2σσσ| १^ / σ|2)   व्य ≡∂ |2
2σσσ|2σσσ|σ|2   2σσσ|2σσσ|σ|2   2σσσ|2σσσ|σ|2

=≡∂| १ / (2σσσ| १ / σ|2)   व्य ≡∂ |2
2σσσ|2σσσ|σ|2   2σσσ|2σσσ|σ|2

=≡∂| १^ / (2σσσ| १ / σ|2)
2σσσ|2σσσ|σ|2 [10 (GSK); II, p.883].

व्य represents subtraction (*vyaya*) of second expression from first one and स represents addition(*saṅkalana*) of the expressions. With this explanation the expressions below: will read in modern notation (replacing ≡ by U) as:

$$\frac{U.\partial.(2\sigma\sigma\sigma).(\sigma).(2)}{(2\sigma\sigma\sigma).(2\sigma\sigma\sigma).(\sigma).(2)} - \frac{\partial.(2)}{(2\sigma\sigma\sigma)\,.\,(2\sigma\sigma\sigma)\,.\,(\sigma)\,.(2)}$$
$$+ \frac{\partial.\{2\sigma\sigma\sigma(2\sigma-1)+1\}}{(2\sigma\sigma\sigma)\,.\,(2\sigma\sigma\sigma)\,.\,(\sigma)\,.(2)}$$
$$= \frac{U.\partial.\,(2\sigma\sigma\sigma)\,.\,(\sigma)\,.(2)}{(2\sigma\sigma\sigma)\,.\,(2\sigma\sigma\sigma)\,.\,(\sigma)\,.(2)} + \frac{\partial.\{2\sigma\sigma\sigma(2\sigma+1)+1\}}{(2\sigma\sigma\sigma)\,.\,(2\sigma\sigma\sigma)\,.\,(\sigma)\,.(2)}$$
$$- \frac{\partial.(2)}{(2\sigma\sigma\sigma)\,.\,(2\sigma\sigma\sigma)\,.\,(\sigma)\,.(2)}$$
$$= \frac{\partial.\{2\sigma\sigma\sigma(2\sigma+1)+1\}}{(2\sigma\sigma\sigma)\,.\,(2\sigma\sigma\sigma)\,.\,(\sigma)\,.(2)} - \frac{\partial.(2)}{(2\sigma\sigma\sigma)\,.\,(2\sigma\sigma\sigma)\,.\,(\sigma)\,.(2)}$$
$$= \frac{\partial.\{2\sigma\sigma\sigma(2\sigma+1)-1\}}{(2\sigma\sigma\sigma)\,.\,(2\sigma\sigma\sigma)\,.\,(\sigma)\,.(2)}.\text{[GSK; p.883].}$$

The above can, therefore, be regarded as beginning of modern algebra in India although the notation was very cumbersome. One will wonder as to why the common factors in numerator and denominator are not cancelled. The reason is this that σ is a variable quantity and each σ may differ in value from the other. However, the value of all the denominators is same.

19. Conclusion.

We have given a panorama of evolution and development of Jaina mathematics over a period from 3rd C. BCE to 14th C. CE We have also shown that some concepts quite advanced for their times such as partition of numbers, occupancy problem of probability theory were definitely evolved before 4th C. CE They were embellished further up to 14th C. CE In that process the greatest contribution was by *Vīrasena* who developed the full basic theory of binomials, logarithms and discussed about pre-calculus concepts like $ln2$ etc. Probably we may trace its influence on Kerala mathematics of India which flourished between 14th and 16th C. as *Vīrasena* operated from just contiguous area. His method of calculation of the volume of conical Universe has great similarity with Archimedes' method of exhaustion.

Acknowledgements

I am grateful to Prof. Anupam Jain for motivation in selecting the theme of the paper, Prof. Surender Jain for his continuous persuation in improving the paper, and finally the reviewers for their numerous suggestions to make the paper flawless.

Abbreviations

A: *Ācārya* å: *aṇgula*
C: Circumference of a circle c: chord of the segment of a circle
D, Dig.: *Digambara* sect
d: diameter of a circle $\bar{d}$: *dhanu*, a linear measure ($=96\mathring{a}$) g: *gāthā*
Ed. Editor, Edn Edition
H, h: height of a frustum of cone or of a segment of a circle
I: infinite
k: *kāṇḍaka*, *kosa*, a linear measure (4 *kosa*s make one *yojana*)
Tr Translator

s: arc of segment of a circle σ: *saṅkhyāta*
sū: sutra $\acute{S}$: *Śvetāmbara* sect
y: *yojana*, a linear measure (=768000$\mathring{a}$)

References

[1] (AB): Āryabhāṭīya of Āryabhaṭa; Āryabhṭa (Hindi) by Gunakar Muley; Information and Broadcasting Ministry, new Delhi, 2nd Ed., 2007.

[2] (ADS): Anuyogadvārasūtra: Commentary in Hindi and Gujrathi by Ghasilalji Maharaja; Akhila Bharatiya svetambara Sthanakvasi jain Shastroddharaka Samiti, Rajkot, 1968.

[2a] (AV): R. S. Shah: Ṣaṭasthānapatita (six-stage) growth (vṛddhī)/ decay (hānī): Arhat-vacana, vol. 29, no.s 3–4, p. 85–102, Indore, India.

[3] (BKS): (i) Bṛhatkṣetrasamāsa of Jinabhadragaṇi Kṣamāśramaṇa with commentary in Sanskrita by Malayagiri; Shri Jainadharma Prasar Sabha, Bhavanagar, 1921. (ii) Gujrathi translation by Vijay Acharya and Nityanand Vijay; vol. I and II; Sanghavi Ambalal Ratanchand Jain Dharmik Trust, Cambay, 1979.

[4] (BS) Vyākhyāprajñaptisūtra, vol. s I to IV: Ed.s Amarmuni andsrichanda Surana 'Sarasa'; Śri Āgama Prakaśana Samiti, Byavara, rajasthan, 2nd Edn., 1991, 1993, 1994, 1994.

[5] (CoP) Crest of Peacock - Non-European Roots of Mathematics: by George Gheverghese Joseph; Penguin Books, 1991.

[6] (Dh) Dhavala of Vīrasena, a commentary on Ṣaṭkaṇḍāgama, Books 1 to 16: Ed.s Hiralal Jain and Phulchanda Siddhantashastri; Jain Sahityoddharak Fund Karyalaya Amaravati, 1938–1958.

[7] (GB1) R. S. Shah: Mathematics of Anuyogadvarasutra; Gaṇita-Bhāratī, Bulletin of Indian Society for History of Mathematics, vol. 29 (2007), no. 1-2, p. 81–100; MD Publications Pvt. Ltd., New Delhi.

[8] (GB2) R. S. Shah: Mathematical Ideas in Bhagavatīsūtra: Gaṇita-Bhāratī: vol. 30 (2008), no.1, p. 1–25.

[9] (GSJ) Gommaṭasāra - Jīvakāṇḍa of Nemicandra Siddhānta-cakravarti, Vols. 1-2, with commentary Samyagdarśanacandrikā by Pt. Ṭoḍaramala: (Hindi): Ed.s Ujjvala Shah and Dineshbhai Shah; Veetaragvaniprakashak, Mumbai, 2007.

[10] (GSK) Gommaṭasāra - Karmakāṇḍa of Nemicandra Siddhānta-cakravarti, Vols. 1–2, with Commentary Samygjñāna-candrikā of Pt. Ṭoḍaramala: (Hindi): Ed.s Ujjvala Shah and Dineshbhai Shah; Veetaragvaniprakashak, Mumbai, 2009.

[11] (IPT) Introduction to Probability Theory, vol. 1, by William Feller; Wiley, 3rd Edn., 2008.

[12] (JD) Jayadhavala of Vīrasena and Jinasena, a commentary on Kaṣāyapāhuḍa, Books 1-16; Ed.s and Publishers as in 6) above, 2nd Edn, 1990–2004.

[13] (JDP) Jambūdvīpaprjñapti Sūtra; Ed./Tr. (Hindi) Chhaganlal Shastri and Mahendrakumar Rankavat; Sri Akhila Bharatiya Sudharma Jain Sanskriti Sangh, Byavara, rajasthan, 2nd Edn. 2006.

[14] (JSK) Jainendra Siddhānta Kośa, vol. 1-4, (Hindi) by Jinendra Varṇī, Bhāratīya Jñānapīṭha Prakāśana, New Delhi, 2070, 2071, 2072, 2073.

[15] (KP) Kaṣāyapāhuḍa of Guṇadhara, (Hindi): ED./TR. Pt. S. C. Diwaker; Śruta Bhaṇḍāra aura Grantha Prakāśana Samiti, Phaltan (Maharashtra).

[16] (Kp) Kammapayaḍī of Śivaśarma, vol. 1-3, (Gujrathi): Ed./Tr. Kailaschanda Vijayji; Śrī. Śvetāmbara Mūrtipūjaka Saṅgha, Surat (Gujrath).

[17] Labdhisāra (including Kṣapaṇāsāra) of Nemicandra Siddhānta-cakravarti (Mādhavacandra Traividya), with commentary Samya-gjñānacandrikā by Pt. Ṭoḍaramala: Tr./Ed. Pt. Phoola-candra Siddhantashastri.

[18] (LV) Lokavibhāga of Sarvanandī (saṃskṛta rendering by simhati-lakasūri): Tr./Ed. Balachandra Shastrī; Jain Sanskriti Samrakshaka Sangha, Solapur (Maharashtra), 1962.

[19] (MB) Mahābandha of Bhūtabalī, 6th Khaṇḍa (volume) of Ṣaṭkhaṇḍāgama, printed separately in 7 Books: Tr./Ed. Pt. Phoolchandra Siddhantashastri; Bhāratīya Jñānapīṭha Prakāśana, New Delhi, 1952–1958.

[19a] (PKB) Pañcakalpabhāṣya, a future publication by JVB, Ladnun, Rajasthan, India.

[19b] (SPS) Pañcasaṅgraha of Candrarṣi, vol. 1-10 (Hindi): Tr./Ed. Devkumar Jain and Misrimalji Maharaj; Ācārya Śrī Raghunātha Jain Śodha Saṃsthāna, Jodhpur (Rajasthan), (1985–1986).

[20] (RV) (Tattvārtha) Rājavārtika of Aklaṅka, vol. 1-2: Tr./Ed. Mahen-drakumar Jain; Bhāratīya Jñānapīḍha Prakāśana, New Delhi, 1953,1958.

[21] (SK) Ṣaṭkhaṇḍāgama, included in 6) above.

[22] (SP/CP) Sūrapaṇṇatti (including Candapaṇṇatti) with Saṃskṛta commentary by Malayagiri: Tr. In Hindi/Ed. Amolak Ṛṣi; Jain Śāstroddhāraka Mudrālaya, Secunderabad, Telangana.

[23] (TP) Tiloyapaṇṇatti of Yativṛṣabha, vol. 1 and 2: Tr. In Hindi/Ed. Adinath Upadhye and Hiralal Jain; Jain Sanskriti Samrakshaka Sangha, Solapur (Maharashtra), 1956 (2nd Edn.), 1951 (1st Edn.).

[24] (TS) Trilokasāra of Nemicandra: Tr. in Hindi /Ed. Pt. Manoharlal Shastri; Hindi Jain Sahityaprasarak Karyalaya, Mumbai, 1918.

[25] (UHN) Universal History of Numbers by Georges Ifrah; John Wiley, 2000.

https://doi.org/10.1142/9789811255502_0003

Article 3

Geometry in Ancient Jaina Works; a Review

S.G. Dani

UM-DAE Centre for Excellence in Basic Sciences
University Campus, Kalina, Santacruz
Mumbai 400098, India
shrigodani@cbs.ac.in

In the context of their pursuit of cosmography the ancient Jaina scholars enunciated various geometric ideas in their compositions. After a lull for a period, in later centuries of the first millennium the earlier geometric understanding was vigorously brought forth and improved upon by various scholars, including Śrīdhara, Vīrasēna, and Mahāvīra, and still later by Ṭhakkura Pherū.

Many formulae from the ancient Jaina compositions have indeed been recalled, in the broader context of study of Jaina works, by various authors, starting with Bibhutibhushan Datta's 1929 paper. The aim of this article is to discuss some of the crucial formulae, and analyse their significance from a mathematical point of view, placing them in the global context. An attempt is also made to place the material in its natural setting, rather than looking at it purely through the prism of present day mathematics.

Keywords: Ancient Jaina Geometry, Formulae for arcs and chords, āyatavṛtta, Śrīdhara, Vīrasēna, Mahāvīra, Ṭhakkura Pherū.

Mathematics Subject Classification 2020: 01-06, 01A32, 01A99.

3.1. Introduction

Various ancient Jaina canonical works are found to have, embedded within them, many mathematical ideas. Geometry in particular was involved in this since the early times, in the cognition of the cosmographical model they adopted.[1] *Sthānānga sūtra*, a canonical work, estimated to be from 300 BCE or earlier ([5], p. 119) gives a systematic classification of the mathematical topics studied, which includes "*rajju*", a term referring to planar geometry, and "*rāsi*", dealing with some aspects of solid geometry; see [18], pp. 67–70. Other ancient works *Candraprajñapti*, *Sūryaprajñapti*, *Jambudvīpaprajñapti*, and others, also devote sections to exposition of geometric principles. The works of *Umāsvāti*, estimated to be from between the 2nd and 4th centuries CE (see [19], p. 20), *Tatvārthādhigama sūtrabhāṣya* and *Jambudvīpasamāsa* stand out as convenient references among the sources of ancient geometric knowledge in the Jaina tradition.

Our current knowledge concerning the period of many ancient works involves considerable uncertainty, and in many cases even the knowledge of their contents is itself based on the commentaries by later scholars, the originals being no longer extant; see [19] for information on available texts. Given this context, our focus here will be on a critical appreciation of the available material from a mathematical point of view, within the historical framework associated with it, and while we shall endeavor, to the extent possible, to indicate the relevant dating of the individual segments of mathematical development, the exposition will be meant to be tentative in this respect.

After a lull for some centuries, near the end of the first millennium the earlier geometric understanding was built upon and carried forward by various scholars, including Śrīdhara, Vīrasēna, Mahāvīra, Nemicandra, Ṭhakkura Pherū, and others. For this later period we are relatively on firmer grounds with regard to the historical specificity, compared to the older period, though some uncertainties remain. Works in this later period also seem to have played an important role in inducting mathematics into various activities including commerce and artisanship, thereby popularizing it. They

[1]A similar model was involved also in the *purāṇic* tradition, but mathematical ideas discussed in the sequel seem to be unique to Jaina literature on the topic.

also made an impact in terms of pedagogy, introducing mathematical ideas to a wider populace. Our aim in this article will be to bring out the significance of the contents of the works, their overall impact, their priority in the overall historical context and influence on other traditions, etc.

3.2. Geometry in Early Jaina Works

Inspiration for study of geometry seems to have been intricately connected with engagements with the model of the cosmos that was envisioned. We briefly recall here the model, to set the context.

Jambudvīpa (= Earth) was visualized as a flat disc of diameter 100000 *yojana*, surrounded alternately by annular rings (*mandalas*), of water and land, of sizes doubling with each ring. The Sun and Moon move in concentric circles in a plane around the earth. The universe was envisioned to consists of three trapezoids piled over one another, with the top and bottom sides of the middle one matching with the top and bottom sides of the ones below and over it respectively.

Engagement with shapes in this way led to introduction of various basic geometric notions. Here are some geometric notions mentioned in *Sūryaprajñapti* (see [28], p. 62, [18], p. 148):

Rectilinear figures: *Trikoṇa* (triangle), *catuṣkoṇa* (quadrilateral), *pañcakoṇa* (pentagon). The reader may note that these terms are also in use in contemporary mathematics in many Indian languages. The term *koṇa*, for angle, is also found in *Sūryaprajñapti*; it may be noted however that numeration of angles, such as in degrees, minutes etc. is not found here. The term *samacatuṣkoṇa* is used for rectangle, and the square was called *samacaturaśra*. Oblique versions of these regular figures were referred to with a prefix *viṣama*, e.g. *viṣamacatuṣkoṇa* stood for parallelogram.

Curved figures: The circle was a frequent occurrence, and was called *cakravāla*. The semicircles were referred to as *cakrārdha cakravāla*. There is also mention of *viṣamacakravāla* which has generally been interpreted in current literature as ellipse, but such precise association with the latter figure seems to be of doubtful validity; the term which translates as "oblique circle" is more likely to have stood for a general figure which is an oblique version of the circle in some, rather nonspecific, sense. No evidence is seen in the literature, justifying the interpretation as "ellipse" either in terms of it being realized as a conic section or in terms of its being the locus of a point whose distances from a pair of points (a posteriori the foci) add up to a constant. Thus the allusion to the concept of the ellipse in the Greek tradition, whose origin is traced to Menaechmus, is problematic; I shall also dwell on this point again when discussing the work of Mahāvīra.

Through the rest of this section we discuss various aspects of the geometry of the circle in the ancient Jaina tradition.

3.2.1. *Circumference of a circle*

One of the notable features of early Jaina geometry is the adoption of $\sqrt{10}$ as the factor involved in obtaining the circumference of a circle from its diameter, at an early stage.

Reckoning the size of the circumference of a circle in terms of its diameter has engaged human civilizations since the early times of development. In all early traditions, including the Chinese, Egyptian, Biblical, as well as in the early Vedic culture in India, the factor involved was taken to be 3. Thus the circumference was understood to be 3 times the diameter, and the same is seen to have been the case in early Jaina tradition; *Sūryaprajñapti* enumerates the circumferences of the *mandalas* around the *Jambudvīpa* applying the traditional factor of 3, but discards them (see [28], page 62), favoring the factor being $\sqrt{10}$.[2]

The factor $\sqrt{10}$ here corresponds to an approximation to π in the modern context, which is about 3.1623, in place of 3.1416...;

[2]In the Vedic *Śulvasūtras* a revision, to $3\frac{1}{5}$, first appeared in Mānava śulvasūtra (in 7th century BCE, which antedates the Jainas); see [2], [3] for the details. The value is however not found used in the *Śulva* tradition. Other than Mānava's sūtra proposing that value as above, the only other place where the factor is involved in the extant *Sūlvasūtras* is a passing reference in *Baudhāyana Śulvasūtra*, where it is taken to be 3; see [3].

thus the error involved is less than 0.7%. The same factor $\sqrt{10}$ was adopted by Chang Heng (also written as Zhang Heng, 78–139 CE), correcting the value 3 that was then prevalent in China. There was considerable contact between India and China during the immediate preceding period, and it may be surmised that adoption of the value in China may have its origin in India.[3]

In India the formula was also used in the Hindu, or *Siddhānta* tradition, by Brahmagupta and others. A more accurate value 3.1416 was described by Āryabhaṭa in *Āryabhaṭīya* (499 CE) but it did not turn up in common usage. In the overall historical context $\sqrt{10}$ is sometimes referred to as the "Jaina value" of π. It continued to be in use as late as the 15th century; see [5], page 131. Along with the Indian arithmetic and astronomy, this value of π also got incorporated in to Arabic mathematics; in particular it is found in Al-Khwarismi's work and is noted to have been from Indian sources; (see [23], page 166).

In various Jaina works (see [5], p. 132 for details) the circumference of *Jambudvīpa* is described, to quite an accuracy, to be 316227 *yojana*, 3 *gavyuti*, 128 *dhanu*, 131 *aṅgula* and a little over. Here *gavyuti*, *dhanu* and *aṅgula* are smaller units of distance prevailing at the time.[4] One *gavyuti* was equivalent to a quarter *yojana*, a *dhanu* was a 2000th part of a *gavyuti*, and an *aṅgula* was 96th part of a *dhanu*; thus 1 *yojana* = 4 *gavyuti* = 8000 *dhanu* = 768000 *aṅgula*. It was apparently computed as the square root of $10 \times (100000)^2 = (10)^{11}$ in *yojanas*. The mention of "a little over", following such an accurate value, is notable; while knowing that to be so would of course be a natural outcome of the computational process by which it was arrived at, that it was considered worth recording shows the importance that they associated with the calculation. In *Tiloyapaṇṇattī* (composed by *Yativṛṣabha*, who is believed to have

[3]Bibhutibhushan Datta recalls, in [5], that (Yoshio) Mikami has stated, in (the first edition of) [22] (p. 70), that "the value $\sqrt{10}$ is found recorded in a Chinese work before it appeared in any Indian work." and proceeds to assert it is incorrect.
[4]The terms *krośa* and *daṇḍa* are used in certain works in place of *gavyuti* and *dhanu* respectively.

flourished sometime between 4th and 7th centuries) the circumference is described to an even finer unit of length, called *avasannāsanna skandha*, which is 8×10^{12}th part of an aṅgula!; see [9] for details.

The square root was apparently computed using the approximate formula $\sqrt{a^2+r} \approx a+\frac{r}{2a}$, where a and r are positive numbers (see [9] for a detailed discussion on this), which is generally known after Heron of Alexandria (ca. 10 – 70 CE); the formula typically was used to find the square root of an integer, say x, by writing it as a^2+r, where a is the largest square integer not exceeding x, and using the expression as above; the formula is valid for r negative as well, but that seems to have been rarely used in ancient times. The formula was known to the ancient Babylonians as well (see, for instance, [7]), but no square root of such a large number seems to have been determined to comparable accuracy prior to this instance. Analogous computations are also found reported in other ancient Jaina works (see [14] for numerous examples).

3.2.2. *On the choice of the value* $\sqrt{10}$

The choice of a factor in the form of a square root is rather unique. Generally in ancient traditions there has been an inclination to choose a fractional number for such factors. Proficiency acquired in computing square roots, may have been a factor involved in adopting a square root for the factor.

We do not know how the factor was arrived at. Various methods have been suggested in literature in this respect. One of these goes back to a commentator Mādhavacandra Traividya (ca. 1000 CE), in his commentary on *Tiloyasāra* of Nemicandra (ca. 975 CE). There are also suggestions coming from later authors, in particular by K. Hunrath in the 19th century (see [28], p. 65), G. Chakravarti (1934) and R.C. Gupta (1986) (see [10]). While Hunrath's argument involves comparison with an inscribed 12-sided regular polygon, the later arguments base themselves on consideration of (inscribed) octagons. The arguments however are mostly unsatisfactory. As explained in [10] the method consists of determining the perimeter of a regular inscribed octagon in a circle with unit diameter, but in the course of the computation involving square roots, approximations by rational numbers are used, causing substantial change in the value, which leads to an answer that is near to $\sqrt{10}$, but still off from it by quite a margin; and yet this is taken as a justification for the choice of the number as the circumference of the circle; it may also be noted, as done in [10] that the perimeter of the regular octagon is substantially less than the circumference, while $\sqrt{10}$ is significantly greater (by over 0.6%). There is also nothing natural about the approximations adopted, that one may grant it as being likely to be picked also by others, in particular the ancient mathematicians. The reasoning of Hunrath described in [28] is also open to similar criticism, to quite an extent, if not fully. In [10] Gupta introduces another possibility, involving a process of averaging. The author mentions that "The process of averaging is known to be a popular and useful ancient technique especially when the exact result of derivation was unknown or difficult". He then proceeds to compute the perimeters of two regular octagons, one inscribed and another circumscribing the circle of unit diameter, and observing that the square of the former is greater than 9 and that of the latter is less than 11, makes a case for $\sqrt{10}$ as the choice for the circumference of the circle. In the present author's view, while the idea of involving the process of averaging may indeed be of significance, too much leeway is taken here in trying to fit the answer; firstly, the actual computations produce values of the perimeters themselves, but averaging is done only of bounds on their squares, without justification for such a step, and secondly the lower and upper bounds 9 and 11 as above are not the precise bounds obtained and thus the "average" is not anything of intrinsic significance. Thus while there is something to the idea of involving the averaging process for an explanation, many questions remain unanswered, in making it a plausible explanation.

3.2.3. *Chords and arrows*

Consider a circle with diameter d and a chord of the circle. The chord divides the circle into two arcs and the chord together with the smaller of the arcs resembles a bow figure. In an obvious extension of the imagery the straight line segment joining the midpoint of the chord to the midpoint of the arc is called the arrow corresponding to the chord. We shall denote by c a chord and by h the corresponding arrow; (the terms stand for the geometric form as well as the associated numerical size, in some fixed units). Various early works contain the following formula relating these quantities[5]:

$$c = \sqrt{4h(d-h)}. \tag{1}$$

This appears in Verse 180 of *Jyotiṣkaraṇḍaka* which purports to expound the knowledge contained in *Sūryaprajñapti* (see [28], p. 63;[6] it may be pointed out here that in a footnote on this page, it is also noted that the same verse, with a modification in the last line, appears in Bhāskara I's commentary on *Āryabhaṭīya*). The formula is also mentioned in Umāsvāti's works *Tattvārthādhigama sūtrabhāṣya* and *Jambudvīpasamāsa*; see [5], pages 124–125.

The following formulae which are deducible from (1) by simple algebraic manipulations have also been noted in the sources recalled above.[7]

$$h = \frac{1}{2}(d - \sqrt{d^2 - c^2}) \tag{2}$$

---

[5]It may be emphasized that while we describe various formulae here using algebraic symbolism, for convenience of exposition, as has been the practice in the recent literature on the topic, the original descriptions are in terms of names of the entities and verbal presentation of the operations involved.

[6]It appears that [28] cites the former for want of access to the text of the latter — relating to the former, [28] mentions in the Bibliograpy, on page 268, a work "*Jyotiṣkaraṇḍaka* of *Vallabhāchārya* with *Malaygiri*'s commentary, Rishabhdevji Kesarimalji Samstha, Ratlam, 1928"; the book however does not seem to be accessible any longer (private communication from Anupam Jain). With regard to the period of *Jyotiṣkaraṇḍaka* I may recall here a passing mention in [28], page 70: "alleged to have been codified at the Valabhi council of the 4th or the 6th century."

[7]By (1) we have $c^2 = 4(dh - h^2) = 4dh - 4h^2 = d^2 - (2h-d)^2$, and equality of the first term with the third and fourth terms leads to (3) and (2) respectively. Though the algebraic operations and the symbolic notation involved here were unavailable in the ancient times, with some practice the calculations involved here can indeed be performed mentally, or with the aid of simple geometric constructions, which is presumably how it would have been done.

and

$$d = \left(h^2 + \frac{c^2}{4}\right)/h. \tag{3}$$

In turn, either of these formulae readily implies Formula 1.

While there has been a general appreciation of the ancients having discovered the formulae, there does not seem to have been a detailed discussion in literature on specifically how they may have arrived at the formulae. Here we propose two possible routes which may have led them to the (interrelated) set of the three formulae. One way of deducing them would be as follows.

Let AD be a semi-chord corresponding to the given chord, and BC be the diameter of the circle passing through D, perpendicular to the chord (see Figure 3.1). Then, as the angle subtended by the diameter at a point on the circumference of the circle, the angle BAC is a right angle. As BC and AD are perpendicular to each other it can be deduced from this that the triangles ABD and CAD are similar triangles. Since the corresponding sides of similar triangles are proportional, it follows that the proportion BD:AD is the same as AD:DC, which is the desired result in Formula 1, as AD, DC and AC are $c/2$, h and $d - h$ respectively.

Before discussing another possibility we observe the following. In the three formulae as above the manifest objective has been to express each of the three quantities, the chord c, the arrow h and the diameter d in terms of the other two, so that a potential user can

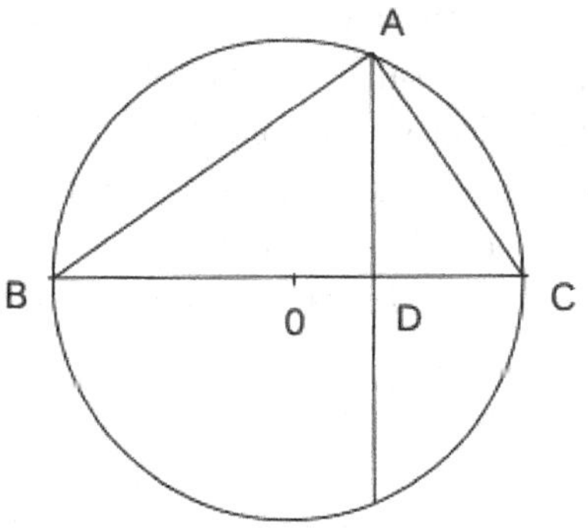

Figure 3.1. Chord and arc

determine the third one, upon knowing any two of them.[8] Underlying these is the quadratic relation

$$\left(\frac{d}{2}\right)^2 = \left(\frac{c}{2}\right)^2 + \left(\frac{d}{2} - h\right)^2. \tag{4}$$

The latter is simply the relation $OA^2 = AD^2 + OD^2$ corresponding exactly to Pythagoras' theorem applied to the right angled triangle OAD; (note that OD is the difference of OC and DC). While the Pythagoras theorem is not found explicitly stated in the extant ancient Jaina works, there are good grounds to believe that they were familiar with it (see below for more on this). This suggests another possibility that Formula (4) could have been noted first, by application of the theorem, and the expressions in the first three formulae, meant for computation of the individual entities in terms of the other two, would have been deduced from it by simple manipulations. Which route they may have followed, it would be difficult to ascertain, but to the present author the second possibility seems more natural and likely in the overall context. If it turns out that it is the first method that was adopted, the discussion above shows how the arguments could have led them to conclude the validity of the Pythagoras theorem in their context (namely obtain a proof for it in the realm of visual geometry, though not in an axiomatic formalism as in Euclidean geometry).

3.2.4. *Chords and arcs*

Now let c be a chord in a circle of diameter d, as before, and let a be the smaller arc segment cut out by the chord from the circumference. What is the relation between (the lengths) c and a? An intriguing formula for this is found along with the formulae as in the last subsection, in the sources mentioned there. Given c and a as above and h the corresponding arrow, the prescription is:

$$a = \sqrt{6h^2 + c^2}. \tag{5}$$

---

[8]The author would have been aware of the mutual relationship between the formulae; in the historical literature on the topic the expressions are listed as if they are independent formulae, which is rather misleading.

Variations of this, describing h and c in terms of the remaining two quantities are also given in *Jyotiṣkaraṅḍaka*, as

$$h = \sqrt{(a^2 - c^2)/6} \text{ and } c = \sqrt{a^2 - 6h^2}$$

(see [28], p. 63).

The reasoning involved in arriving at such a formula seems to be broadly the following: Consider the triangle-like region (see Figure 3.1) between the semi-chord AD, the arrow DC and the half the arc-segment joining A to C; it resembles a right angled triangle, except for the diagonal segment being curved. The length of the straight segment AC, joining one of the endpoints to the midpoint of the arc, is given by the Pythagoras theorem to be $\sqrt{h^2 + (c/2)^2} = {}^1\!/\!_2\sqrt{4h^2 + c^2}$. On account of being curved the arc segment between the midpoint and the endpoint has to be greater than that. Taking into account the similarity of the situation with the right angled triangle ADC and the Pythagoras theorem one may seek an expression in the form: ${}^1\!/\!_2\sqrt{sh^2 + c^2}$, with a multiplier $s > 4$ in place of 4. We see that for the expression to tally in the case when the chord is the diameter (so $c = d$ and $h = \frac{1}{2}d$, in which case $a = \frac{1}{4}\pi d$) we should have $\sqrt{\frac{1}{4}s + 1} = \frac{1}{2}\pi$; since π was taken to be $\sqrt{10}$ this leads to $s = 6$, which was chosen constant.[9]

The issue of relating the length of an arc to the corresponding chord was also involved in the work of Heron of Alexandria (ca. 10–70 CE). He introduced two formulae:

[9]We may also recall here the following: a formula for the semi-arc in terms of the half-chord (sine) and the arrow (versine) is described by *Nīlakaṇṭha Somayājin* in his commentary on *Āryabhaṭīya*, which in the notation as above corresponds to:

$$\frac{a}{2} = \sqrt{\frac{4}{3}h^2 + \left(\frac{c}{2}\right)^2}, \text{ or } a = \sqrt{\frac{16}{3}h^2 + c^2};$$

the formula is derived using 'infinitesimal' methods and is recommended, by *Nīlakaṇṭha* to be used only for small arcs; (see [28], pp. 179–182 for details). It is also mentioned in [28] (page 182), interestingly, that the commentator states that the derivation "is implied" by the second part of *sūtra* 17 of *Gaṇitapāda* of *Āryabhaṭīya* (following the statement of the 'Pythagoras thorem'); linguistically however that part actually corresponds to Formula (1) described in §1.3 above, relating the diameter and the arrow; see [31], page 59.

The first one is $a = \sqrt{4h^2 + c^2} + {}^1\!/_4 h$ (in the notation as above, making a simple linear increment in terms of h in the term $\sqrt{4h^2 + c^2}$ corresponding to the length of the line segment path noted above). The second one, which is apparently meant to be a more refined one, is

$$\sqrt{4h^2 + c^2} + \{\sqrt{4h^2 + c^2 - c}\}\frac{h}{c} = \sqrt{4h^2 + c^2}\left(1 + \frac{h}{c}\right) - h$$

([16], p. 331, and [5], p. 130). It is also mentioned in [5] (p. 130), citing (an earlier edition of) [22], p. 62, that Ch'en Huo (in China, 11th century CE) gives the formula $a = c + 2h^2/d$.

These formulae are significant as the earliest attempts, in the pre-trigonometry era, towards understanding the interrelations between the chord, the arrow, and the corresponding arc.[10] In the following table we list the values obtained from the formulae as above, together with the true values, upto 4 decimal digits, for the arclength corresponding to chords in the unit circle subtending an angle 2θ (viewed for convenience as the "vertical" chord subtending the angle from $-\theta$ to θ with respect to the usual coordinatization) for various values of θ; we note that in this case $c = 2\sin\theta$ and $h = (1 - \cos\theta) = 2\sin^2(\frac{1}{2}\theta)$, and hence the expressions involved in the Jaina and Heron's formulae are given, respectively, by the following[11]:

$$\sqrt{6h^2 + c^2} = 2\sin(\theta/2)\sqrt{5 - \cos\theta},$$

$$\sqrt{4h^2 + c^2}\frac{1}{4}h = 4\sin(\theta/2) + \frac{1}{2}\sin^2(\theta/2), \text{ and}$$

[10]We note also the following: In the *Siddhānta* astronomy there is a well-known approximate formula, with considerable accuracy, for $\sin\theta$, known after Bhāskara the first (though it is found also in Brahmagupta's work, and may also have been known earlier); see [8] for details. The formula expresses $\sin\theta$ as a rational function of θ. The function is a ratio of two quadratic polynomials and may be readily be inverted to produce a formula for θ in terms of $\sin\theta$, which corresponds to the issue at hand. We shall however not go into the details here.

[11]We have $6h^2 + c^2 = 6(1 - \cos\theta)^2 + 4\sin\theta^2 = 2(5 - 6\cos\theta + \cos^2\theta) = 2(1 - \cos\theta)(5 - \cos\theta) = 4\sin^2(\theta/2)(5 - \cos\theta)$, which gives the first equality as above. Also $4h^2 + c^2 = 4(1 - \cos\theta)^2 + 4\sin^2\theta = 4(2 - 2\cos\theta) = 16\sin^2(\theta/2)$, which readily implies the other two equations.

$$\sqrt{4h^2 + c^2}\left(1 + \frac{h}{c}\right) - h = 4\sin(\theta/2)\left(1 + \frac{1}{2}\tan(\theta/2)\right) - (1 - \cos\theta).$$

As $d = 2$ Ch'en Huo's expression corresponds to $2\sin\theta + (1 - \cos\theta)^2$ in this instance.

Angle	arc-length true value	Jaina value	Heron's 1st value	Heron's 2nd value	Ch'en Huo's value
15°	0.5236	0.5243 (+0.0007)	0.5306 (+0.0070)	0.5224 (−0.0012)	0.5188
30°	1.0472	1.0525 (+0.0053)	1.0688 (+0.0216)	1.04 (−0.0072)	1.0179
45°	1.5708	1.5858 (+0.0150)	1.6049 (+0.0341)	1.5549 (−0.0159)	1.5
60°	2.0944	2.1213 (+0.0269)	2.125 (+0.0306)	2.0774 (−0.0170)	1.9821
75°	2.618	2.6511 (+0.0331)	2.6203 (+0.0023)	2.6281 (+0.0101)	2.4812
90°	3.1416	3.1623	3.0784	3.2426	3

The table shows that the values from the Jaina formula give good estimates to the corresponding true values. In [5] (p. 130) Datta states "It will be observed that the Hindu value of the arc is older and more accurate than the other two." It may firstly be clarified that by "Hindu" value the author means the values from the Jaina tradition as discussed above.[12] Secondly, while Datta mentions "value" in singular, presumably meaning the statement to be true for any chord and arrow, the table shows that the statement regarding comparisons does not hold consistently for all values involved, except with regard to the value of Ch'en Huo. Little is known about the

[12]The issue did not feature, as far as is known, in the *Siddhānta* tradition which may be suggested by the reference; elsewhere in the article, as also at various other places in literature, the term Hindu has been used in a similar fashion, which includes "Jaina" with "Hindu".

background of the latter (see [22] for some details confirming this), and the prescription may have been meant for usage in some practical context — it may be noted that the stipulated computation is much simpler compared to the other formulae, and that the formula implicitly takes the value of π to be 3, whereas much better values were known in China since the times of Liu Hui (3rd century) and Zu Chongzhi (5th century). Also, Ch'en Huo seems to be a relatively minor figure in Chinese mathematics, not commonly mentioned by historians of Chinese mathematics. On the other hand the poor choice involved, as late as the 11th century, could also be related to the fact the mathematical, and general scientific, tradition in China has witnessed major ups and downs over the centuries. On the whole a comparison with values of Ch'en Huo does not seem to be of much relevance. With regard to comparisons with Heron's formula the relative level of accuracy of the two formulae is seen to depend on the range of the angle involved; it is remarkable however that the formula is more accurate for the most part than Heron's first formula, and more elegant than his second refined one.

3.2.5. *Area formulae*

Along with the various formulae concerning various lengths that we discussed with reference to *Jyotiṣkaraṅḍaka*, representing the knowledge from *Sūryaprajñapti*, as also *Tattvārthādhigama sūtrabhāṣya* of *Umāsvāti*, there is also a formula describing the area A of the circle being given by

$$A = \frac{1}{4}Cd, \tag{6}$$

where C is the circumference and d is the diameter of the circle.[13]

[13] In ancient cultures typically the ratio of circumference to the diameter and that of the area of a circle to the square of its radius appear independently, and the corresponding numerical values are often not equal; the equality of the two ratios, which is an a posteriori fact, was historically inferred at some stage, through reasoning, whether analytical or heuristic. In the Jaina context, once the area was realized by such a formula, in terms of the circumference, the equality is automatic.

This relation may have been recognized by thinking of the partition of the circle into thin isosceles triangles with a common vertex at the centre of the circle and the other two on the circumference. Each of these triangles have height $\frac{1}{2}d$ and base a little segment of the circle. The area would have been realized, heuristically, to be the sum of areas of these triangles, thereby deducing that the total area is $\frac{1}{4}Cd$. There however does not seem to be any specific evidence towards this, and a claim for such reasoning involving "infinitesimals" may not be sustainable.

While in the later Jaina works we find formulae for areas of segments of circles cut off by chords, which we shall come to later, the ancient works do not seem to have engaged with such a question (see [5], p. 130). Similarly formulae for volumes, other than for certain simple rectilinear shapes, seem to have made an appearance only later.

3.3. Jaina Geometry in Later Phases

There seems to have been a lull for a while in the interest in mathematics in the early centuries of CE, after Umāsvāti, until around the 8th century when the earlier geometric understanding was vigorously brought forth and carried forward by various scholars.[14] During the period from the 8th to 10th or 11th centuries we come across important works, due to Śrīdhara, Vīrasēna, Mahāvīra and Nemicandra. Again from somewhat later, late 13th and early 14th centuries, we have the work of Ṭhakkura Pherū which made an impact. In this section we shall discuss some aspects of these works, focusing on geometry.

Before going into the mathematical details, a few words putting the development in an historical perspective would be in order. The period in question succeeds the heyday of the Siddhānta tradition of mathematical astronomy, with the works of Āryabhaṭa, Brahmagupta, and Bhāskara the first; the name of Varāhamihira may

[14] A few names and works from the interim period have been mentioned in literature (see [19], p. 20, in particular) but no mathematical details have been prominently known.

also be added here for the broader context, though his contributions relate primarily to astronomy (and astrology) and with no significant mathematical component, unlike in the case of the others mentioned; see [25] for a general reference. Āryabhaṭa's *Āryabhaṭīya*, composed in 499 CE, set the tone, serving as pioneering work, for a tradition that lasted over a thousand years. It was also inspirational to the Kerala school of Mādhava that flourished from the middle of the 14th century for almost 3 centuries, making remarkable contributions. Āryabhaṭa was followed in a little over a century by the works of Bhāskara the first, in the early part of the 7th century, which not only elaborated on Āryabhaṭa's concise presentation and clarified various matters, but also introduced new techniques in astronomical as well as mathematical aspects. Around the same time in 628 CE Brahmagupta composed his *Brāhmasphuṭasiddhānta,* which is an extensive work in the tradition, though critical of Āryabhaṭa in various respects. Both Bhāskara the first and Brahmagupta show influence of the earlier Jaina works in some ways, including the use of $\sqrt{10}$ for π (which is absent in *Āryabhaṭīya*).

As to be expected, the *Siddhānta* tradition did sustain a broad mathematical learning. However its *raison d'etre* remained pinned to astronomy, with applications around astrology and issues involved in producing almanacs, predicting eclipses etc., and the mathematical learning associated with it remained confined to the community engaged with these pursuits. It was not until Bhāskarāchārya's *Līlāvatī* that an independent book on mathematics seems to have emerged in that tradition; *Līlāvatī* and its successor *Bījagaṇita*, were also meant to be parts of the larger treatise *Siddhāntaśiromaṇi* in the then prevailing model of *Siddhānta* works. However by this period the general interest in the mathematical topics discussed in the work had got extended well beyond the community involved with astronomy, as a result of which the *Līlāvatī* part came to be copied numerous times and studied on a much wider scale than the larger treatise, thus giving *Līlāvatī* an identity as a mathematical book on its own. A similar development occurred, though on a smaller scale, with regard to *Bījagaṇita,* which is more technical than *Līlāvatī* but nevertheless of considerable independent interest, outside the ambit of astronomy.

Accumulation of basic mathematical understanding emerging from the Jaina philosophical tradition, the *Siddhānta* tradition of mathematical astronomy, and possibly also imports from other cultures, from within the country and also from outside, seem to have gradually percolated to a wider populace, outside the limited circles of scholars and practitioners engaged with it, creating an appetite for mathematical knowledge. Mathematics by this time seems to have come into greater and active contact with trade, artisanry and a variety of practical activities, and sources of independent learning of mathematics would have been sought after. The Jaina scholars noted above catered substantially to the emerging appetite for mathematical learning, and were also instrumental in popularizing the use of mathematics in a range of areas. They also shaped the mathematical activity of the time with new contributions. Their works inevitably played a pedagogical role, incorporating as they did, various features engendering interest in the subject. We shall now discuss the works of some select authors in respect of some geometric aspects involved in the work. As many of the basic geometric ideas are common to all of them, and have been discussed in current literature especially in the context of the pedagogical role, our emphasis here will be on the features that distinguish the individual works.

3.3.1. *Śrīdhara and the volume of a sphere*

There has been a debate among historians on whether Śrīdhara belongs to the Jaina tradition or not, but thankfully it now seems to be settled, confirming that to be the case. At any rate the overall academic profile of his work is consistent with his being from the Jaina tradition, providing adequate testimony in that respect. Similarly, there had also been some confusion about his period, and in particular whether he preceded or succeeded Mahāvīra. It is now confirmed that he flourished sometime around the middle of the 8th century, *before* Mahāvīra; see [11] for various details in this respect — see also [20] and [32].

Two of his works have come down to us, *Triśatikā* and *Pāṭīgaṇita*, the latter only in a single copy, which is also incomplete. Both the works deal with a variety of topics in arithmetic and geometry.

Pāṭīgaṇita (see [30]) in particular contains a large number of examples and illustrations from everyday life. On account of this feature it also throws light on various aspects of life during the author's time. After a discussion of the formulae for areas of rectangles, trapezia, triangles etc. the circumference and area of the circle are described, with $\sqrt{10}$ as the value for π, following the Jaina tradition. Though Āryabhaṭa had introduced a more accurate value 3.1416 (the original is in terms of the circumference of a circle of diameter 20000, which in the decimal notation corresponds to this value of π), it did not attain currency even in the scholarly circles, and use of the Jaina value was preferred by many authors.

For the area of a quadrilateral Śrīdhara first recalls the crude rule, as the product of the average lengths of the pairs of opposite sides, namely $\frac{a+c}{2}$, $\frac{b+d}{2}$, where a, b, c and d are the sides of the quadrilateral, labeled cyclically. Such an expression for the area goes back to *Barahmagupta's Brāhmasphuṭasiddhānta* (verse XII, 21, where it is referred to as *sthūlaphalaṃ*) and has been repeated in various sources.[15] After recalling the formula Śrīdhara points out in *Pāṭīgaṇita* specifically that it is applicable only when the differences in the sizes are small, and proceeds to a discussion on more exact formulae, giving in particular the formula for the areas of trapezia (see [30]).

In *Triśatikā* the formula $\sqrt{s(s-a)(s-b)(s-c)}$ for the area of a triangle, where a, b, c are the sides and s the semi-perimeter of the triangle, which is generally known after Heron of Alexandria, is described. The corresponding formula $\sqrt{(s-a)(s-b)(s-c)(s-d)}$ where a, b, c, d are the sides and s semi-perimeter of a quadrilateral, known after *Brahmagupta*, is mentioned; the author however misses,

[15] I was struck, and rather dismayed, to find that the overseer involved in the construction of my apartment building also used the formula!

like various other Indian authors around the period,[16] stipulating the condition of cyclicity that is required for its validity.

Let me now come to a notable feature in Śrīdhara's work which does not seem to have received due attention. In the Indian context the credit for having been the first to give the correct formula for the volume of a sphere, is normally given to Bhāskarāchārya (12th c.); see [28], p. 210.[17] However, Śrīdhara's *Triśatikā* (verse 56) describes the volume (in our notation) to be $\frac{d^3}{2}\left(1+\frac{1}{18}\right)$, where d is the diameter; see [12] for a detailed discussion. It would seem that Śrīdhara had the correct formula $\frac{4}{3}\pi\left(\frac{d}{2}\right)^3$ in mind, but for the factor $\frac{\pi}{3}$ he put down only the approximate value $1+\frac{1}{18}$. This ought to be seen in context by recalling that unlike now there was no standard notation for π at that time; as $\sqrt{10}$ was the routinely adopted value for π the factor could have been expressed as $\sqrt{10/9}$, in which case we could have been more certain that he means it to be $\frac{\pi}{3}$. However, as *Triśatikā* was meant to be a short introduction, and the formulae were written down to facilitate computation by potential users of them, a simpler expression in the form of a fraction may have been preferred; the issue of versification of the statement may also have contributed to choosing the format with the fraction, which in this instance is easy to split and describe, compared to $\sqrt{10/9}$. It may be noted in this respect that $1+1/18$ was the standard approximation that time for the factor $\sqrt{10/9}=\sqrt{1+1/9}$ involved (by the square root formula which we alluded to earlier, in Section 3.2.1). Unfortunately as there is no indication of how the volume formula was arrived at, and the reference to the issue is not found in the extant part of Pāṭīgaṇita, it would not be possible to ascertain such a conjecture.

Notwithstanding whether the formula may be identified with the correct one as we now know, it is quite a good formula in

[16]While apparently Brahmagupta meant the formula to be for cyclic quadrilaterals (see [21] for details) it is not adequately clear from the text, which seems to have led to perpetual confusion on the issue in India, until it was refuted by Āryabhaṭa II (11th century) and Bhāskaracārya (12th century); see [4].

[17]It may be recalled in this respect that *Āryabhaṭīya* purports to give a formula, but it is incorrect; see [28], p. 197; other early Indian authors have not discussed the issue.

terms of accuracy, involving an error of less than 1 percent. It is a peculiar quirk of history that despite such a good and usable formula having been discovered, various later authors even in the same tradition, including some celebrated ones, did not adopt it; the formulae described by Mahāvīra about a century later, and even by Pherū as late as the 14th century, are substantially cruder than this formula.

Curiously Śrīdhara does not seem to have considered the surface area of the sphere, which in the Jaina tradition appears in the work of Mahāvīra (see infra); see [12] for details on various volume computations in ancient India. It may be mentioned however that as in classical Jaina mathematics Śrīdhara considers regions cut off from the circle by a chord and the minor arc (the smaller of the arcs resulting from the division); however, the formulae for the arclength etc. are not recalled, but a formula is given for the area of the region between a chord and minor arc, to be $\frac{\sqrt{10}}{3} \cdot \frac{c+h}{2} \cdot h$ (c and h are the chord and arrow respectively, as before); the formula is somewhat crude, and what the general idea in the derivation might have been is not clear.[18] For the same area Mahāvīra in *Gaṇitasārasaṅgraha* gives the expression $\sqrt{10}ch/4$ which is quite crude (Datta describes it as "wrong"; see [5], p. 145). Nemichandra gives the latter value as gross (*sthūla*) value, mentioning Śrīdhara's value as neat (*sūkṣma*) one; see [6].

3.3.2. *Vīrasena and the volume of a conical frustrum*

Vīrasena, the author of *Dhavalā Tīkā* on the *Ṣaṭkhaṇḍāgama* apparently flourished in the eighth or ninth century (the year 816 is listed as his period in [19]; [28] however mentioned him as being from the 8th century). He is often quoted for a good approximation to the value of π given in the verse from *Ṣaṭkhaṇḍāgama,* Vol. IV.

व्यासं षोडशगुणितं षोडशसहितं त्रिरूपरूपैर्भक्तम् ।
व्यासं त्रिगुणितं सूक्ष्मादपि तद्भवेत् सूक्ष्मम् ।

[18]It may be recalled here that *Metrica* of Hero of Alexandria offers $\frac{c+h}{2}\, h + \frac{1}{14}\left(\frac{c}{2}\right)^2$ as the formula in this respect, and the same was adopted in the ancient Hebrew text (period uncertain) Mishnatoha-Middot; see [23], p. 163.

Vyāsaṁ ṣōḍaśaguṇitaṁ ṣōḍaśasahitaṃ trirūparūpairbhaktaṁ |
Vyāsaṁ triguṇitaṁ sūkṣmādapi tadbhavet sūkṣmaṁ |

"The diameter multiplied by 16, together with 16, divided by 113, and three times the diameter becomes finest of fine (value of the perimeter)" (translation as in [28]). The part "together with 16," is a rather puzzling feature in this, since one could not be adding a fixed number independent of the size of the diameter in the computation of the circumference. When the part is dropped out of consideration we see that what is described corresponds to an approximation to π as $3\frac{16}{113} = \frac{355}{113} \approx 3.1415929\ldots$, coinciding with the true value for π upto 6 decimal places, the latter being 3.1415926 .... Such a value was earlier proposed in China by Zu Chongzhi (5th century), who determined the value of π to be between 3.1415926 and 3.1415927 and deduced from it in some way the fractional approximation as above (see [22], p. 50). In the overall context it seems plausible that the value mentioned by Vīrasena may ultimately have its origin in the work of Zu Chongzhi.

That leaves us with the issue of the "together with 16" part mentioned above. Generally the response in the literature on the topic to this peculiarity has been not to pay attention to it. In this regard I would like to record here a suggestion, of linguistic nature, which may explain the point. The suggestion is that "together with 16" relates to the same 16 appearing in the earlier part. Thus a small emendation in the interpretation, somewhat like "The diameter to be multiplied by 16 and the product together with that 16 to be divided by 113 ...", would set things right (thus, while *sahitaṃ* indeed suggests addition, that need not be the only interpretation, especially when it is seen to lead to a manifestly wrong inference); in a scholarly statement one does not normally expect such a repetition of the kind being suggested, but in colloquial communication it seems well imaginable, and hence also may be involved as a valid form in some linguistic practice. Moreover, if it is true that the statement has a Chinese precursor then it is possible the nature of the original statement could have prompted the repetition, even though it is uncalled for and misleading in the Sanskrit rendering.

Vīrasena actually deserves to be better known for another of his work which does not seem to have received much attention; this consists of his formula for the volume of a conical frustrum; see [28], p. 203. In his *Dhavalā Ṭīkā* the volume of such a frustrum with diameters a and b at the base and the top respectively, and height h is stated to be $\frac{\pi h}{4} \cdot \frac{a^2+ab+b^2}{3}$. It is worth mentioning that, unlike in much of the extant ancient and medieval mathematical literature in India, we find in this work a description of the method of determining the volume. Moreover, the method involves summation of an infinite series and the idea of infinitesimals, akin to Calculus; see [28], pages 203–205 for details of the computation. While neither the formula nor the general idea of using infinitesimals as involved here may be considered unprecedented in the global context, and are reminiscent of computations going back to Archimedes and Liu Hui, in the Greek and Chinese traditions respectively, the details are arguably new, and seem to be a notable first, especially in mathematics in India.[19]

3.3.3. *Mahāvīra on the quadrilaterals*

Gaṇitasārasaṅgraha of *Mahāvīra*, from around 850 CE, has been one of the very influential books in mathematics and mathematical education, especially in South India. It served as a textbook of mathematics over a broad geographical region for some centuries, quite likely until the other popular medieval Indian book on arithmetic and geometry, *Līlāvatī*, took the place sometime in the 12th century. *Gaṇitasārasaṅgraha* is an extensive and leisurely exposition of arithmetic, combinatorics, and geometry, with numerous examples in the form of exercises.

Kṣetragaṇitavyavahāraḥ, the Chapter on geometry, is the second largest of the nine chapters in the book, with well over 200 verses, that include a large number of numerical examples. An interesting feature of the exposition is that there is a separate section in the chapter devoted to formulae for practical usage, described as *vyāvahārika*

[19]In particular it seems plausible that it may have influenced Bhāskaracārya, who apparently was familiar with the work of Śrīdhara. It would however take a closer analysis of the two to come to any definitive conclusion.

gaṇitaṃ. Many relations described here are approximate, *sthūla*, in the nature of thumb rules facilitating quick computation; on the other hand they cover various geometric forms not commonly dealt with in other works, e.g. the shape of a conch or elephant tusk etc. Better formulae, often involving somewhat more intricate expressions, are described separately later. These are described as *sūkṣma* (literally meaning "fine") values; they are meant to be the better values as perceived by the author, typically involving a little more intricate computation compared to the *sthūla* values, but not necessarily exact in general; in some instances the formulae from both the groups are only surmised ones (and were apparently not derived) and are inexact.

Mahāvīra also recalls the formula for the area of quadrilaterals that goes back to Brahmagupta, which we recalled earlier. In his Introduction to *Gaṇitasārasaṅgraha,* edited by M. Raṅgācārya [26], David Eugene Smith seems to make a point, on p. xxiii, that it is not observed that the formula holds only for a cyclic figure.[20] In this respect Sarasvati Amma points out ([28], p. 92) that the relevant verse (GSS VII.50) excludes "*viṣama caturarśra*" from application of the formula, and follows it with the comment "which makes it probable that he was aware of the restriction in the formula". Whether this properly clarifies the issue, and a similar point arising in respect of the formula for the diagonals of the quadrilaterals is debatable. We shall however not go into further discussion on it here. Mahāvīra also deals with a variety of problems concerning construction of quadrilaterals, involving ideas from the general areas of geometric algebra and Diophantine equations. A comprehensive analysis and exposition of these parts of *Gaṇitasārasaṅgraha* would be worthwhile, but we shall not go into it here.

3.3.4. *Mahāvīra, on curved figures and the āyatavṛtta*

One of the interesting features of Mahāvīra's *Gaṇitasārasaṅgraha* with regard to geometry is the consideration of issues of mensuration

[20] His observation extends to Brahmagupta as well which, as mentioned earlier, has been clarified in [21].

of certain figures that are generally not found elsewhere in literature. Among these are various planar figures involving combinations on the theme of circle and semicircle: a conchiform area (*kambukāvṛtta*, thought of as formed of two semicircles of different diameters joined along the diameters, on one side of the longer one). Then there are surfaces with spatial curvature. He considered also sections of the sphere cut off by planes, in the form of concave and convex circular areas (called *nimnavṛtta* and *unnatavṛtta* respectively), as well as outward and inward curved annular shapes (*bahiṣcakravālavṛtta* and *antaṣcakravālavṛtta*). While on the one hand this is in the nature of generalization of ideas from ancient Jaina literature, about the interrelation of the arc and the chord, there seems to be a strong motivation here in the form of practical applications; the examples to which the formulae are applied include various interesting practical shapes, such as the tortoise-back or concave sacrificial pit. The formula for the area of such a surface is given, in GSS-VII-25, as $\frac{1}{4}$th of the product of the circumference of the circular boundary and what is referred to as the "*viṣkambha*". The latter term has been interpreted in much of the historical literature on the topic, including [26] and [24], as meaning the diameter of the same circle. Such an interpretation is erroneous, as has been pointed out by R.C. Gupta (see [13]); the same product was given as the gross (*sthūla*) value for the area of the planar disc through the boundary circle; actually they were aware that the latter was smaller than the fine (*sūkṣma*) value for the same figure, and would surely have realized that the area of the curved surface would be even more than that. An interpretation of the term *viṣkambha* as the diameter along the surface, viz. length of the curve on the surface joining diametrically opposite points contained in a plane passing through the centre of the sphere, is more satisfactory, and has been justified in [13] based on a historical perspective on such empirical formulae.

The formula applied in particular to a hemisphere yields the area to be $C^2/8$, where C is the circumference, and thus the area of the sphere may readily be inferred to be $C^2/4$. As recalled earlier in this article Mahāvīra seems to have been the first one in the Jaina tradition to have had a formula (though implicitly) for the surface area of a sphere. The reader may note that the true value of the area of the sphere is C^2/π, so the value here is rather crude, but is of historical significance as an empirical value. The formula was later improved by Pherū to $\frac{10}{9}C^2/4$, perhaps on experimental grounds, bringing it a little closer to the true value; see [27].

Along with these there is also another figure mentioned, which has been called *āyatavṛtta*. The term has been translated in current literature (e.g. [26], [28], [24], [13], and others) on the topic as "ellipse". This however does not seem to be a valid choice, when the latter term is understood in the mathematical sense, originating from the Greek tradition.

Unfortunately the work contains no description of how the figure was drawn, nor any clue about it. The term *āyatavṛtta* literally means stretched or elongated circle, and a variety of geometric shapes would qualify for such a description in terms of an heuristic interpretation of the phrase. As we shall see below the formulae proposed for the circumference are based on specific familiar "models" for "elongated circles", as shown in Figure 3.2 (see details below), neither of which is an ellipse in the usual sense of the term. Thus the identification of the "*āyatavṛtta*" with ellipse lacks proper justification.

We recall that the formulae for circumference are given in two versions, *sthūla* and *sūkṣma* ("gross" and "fine" respectively). They involve two parameters, called *āyāma* and *vyāsa,* corresponding to the longer and shorter (understood to be axial) dimensions, respectively. With a for *āyāma* and b for *vyāsa* the *sthūla* value of the circumference is given to be $2(a + \frac{b}{2})$. This may be readily seen to correspond to the circumference of the "elongated circle" consisting of two semicircular arcs of a circle of diameter b, joined by two straight line segments of length $a - b$, as in the top part of Figure 3.2, with the circumference of the circle taken to be three times the diameter, adopting the *sthūla* value for the factor; the total circumference is then $3 \times b + 2(a - b) = 2(a + \frac{b}{2})$.

For the *sūkṣma* value the author evidently falls back on the traditional Jaina formula for lengths of arcs of circles cut off by chords (discussed in §3.2.4); see also the discussion in [13] in this respect. The *āyatavṛtta* is now visualized as two such identical arcs joined

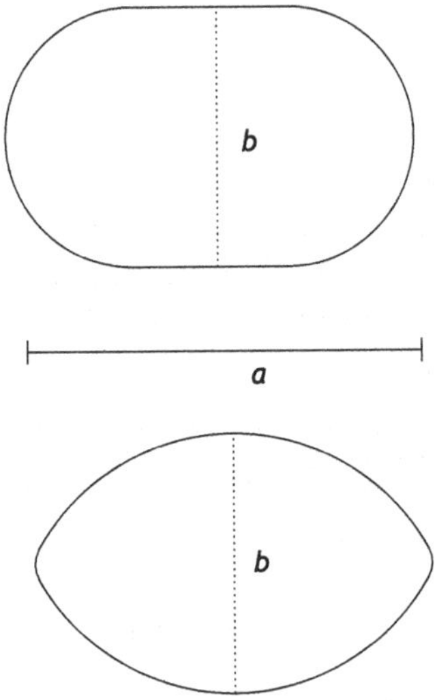

Figure 3.2. Āyatavṛtta

symmetrically (as in the bottom part of Figure 3.2), ignoring the cusps formed at the two ends. For each of the arcs the corresponding chord is the *āyāma* of the *āyatavṛtta*, viz. a as above, and the "arrow" is half of the *vyāsa*, which is $\frac{b}{2}$. Hence by Formula (6) the length of each of them is $\sqrt{\frac{3}{2}b^2 + a^2}$, and the combined length of the two arcs is $\sqrt{4a^2 + 6b^2}$, which is given as the *sūkṣhma* value of the circumference of the *āyatavṛtta.*

The area of the *āyatavṛtta* in either of the *sthūla* and *sūkṣhma* versions is given to be half the product of the respective circumference with what is called "*viṣkambha*"; in most of the current literature, including [24] and [26] the latter term has been interpreted as synonymous with *vyāsa,* while [13] makes a convincing case for a different interpretation. With the modified interpretation the value assigned turns out to be substantially closer to the true value, even though it is still a crude empirical assignment, meant for practical day to day use.

In the Greek tradition the notion of conic sections arose out of the problem of the means, and the notion of an ellipse was a natural outcome of the studies in this respect. In India there was no such specific context for introduction of a geometrical shape that we now call the ellipse, and one does not also find any unambiguous reference to such a notion. The planetary models also did not involve ellipses at that time, but rather were based on epicycles. It would thus be more appropriate to suppose that *āyatavṛtta* was meant to be a more general figure fitting the overall idea of an elongated circle, and the formulae were meant for approximate practical computations, which is very much the spirit of exposition as a whole.

3.3.5. *Ṭhakkura Pherū*

Ṭhakkura Pherū is another Jaina mathematician to have made a significant impact on promotion of mathematical understanding, in North India. He was a polymath, with significant contributions in a variety of branches of knowledge including, Gemmology, Astronomy, Architecture, Metallurgy, besides mathematics, in the court of Khaljī Sultāns of Delhi during the early decades of the 14th century. In mathematics Pherū contributed the *Gaṇitasārakaumudī* (see [27]) which apparently had been very influential during his time. He is also known to have been the first to provide general constructions of magic squares, though there has been much interest in them in India over a long period; see [27]. Unlike the works of Śrīdhara and Mahāvīra, which are in Sanskrit, *Gaṇitasārakaumudī* is in *Apabhraṁṣa Prākṛta.* Perhaps due to this, the work has not been much studied by the modern scholarship. The work [27] has made a good beginning in this respect, by providing a translation and putting the contents in perspective. The new context of the onset of Islamic culture in India during the period is manifest in Pherū's work in various ways. With regard to geometry for instance, computations relating to minars, arches etc. may be mentioned in this respect. Apart from having been influential in introducing mathematical knowledge in various activities, he seems to have been perceptive on issues of accuracy of the (empirical) formulae that were in use and is known to have introduced some corrections as we noted earlier. Hopefully more detailed further studies will throw more light on the specific mathematical significance of the work.

3.4. In Place of Conclusion

Much of the literature on mathematics from the Jaina tradition, as also with various other ancient cultures, has been in the nature of cataloguing various mathematical notions and observations found in the texts, coupled with some claims on historical priority etc. with relatively little focus on aspects of understanding and anticipating the development of ideas from an intrinsic perspective. While emphasis on the first part is inevitable to an extent in the early stages of any developing studies in any historical tradition, there is now a need to direct efforts towards progress in the latter. With this in view we have analyzed various formulae, ideas involved, and concepts introduced. It is hoped that the discussion in the preceding pages will serve as a small step towards such an endeavor.

Acknowledgements

The author would like to thank Professors Anupam Jain and Surender K. Jain, and Mr. Pankajkumar K. Shah for encouragement and help in preparing this paper. Thanks are due to Dr. Qijun Yan and Dr. Lovy Singhal for making available a copy of the book of Yoshio Mikami. Thanks are also due to Dr. Manoj Choudhuri for producing the figures used in the article, and to Mr. Anil Kumar Jain and Mr. Umashankara Kelathaya for help in preparation of the manuscript in word format.

References

[1] Dani, S.G., Geometry in the Śulvasūtras, in Studies in History of Mathematics, Proceedings of Chennai Seminar, Ed. C.S. Seshadri, Hindustan Book Agency, New Delhi, 2010.

[2] Dani, S.G., Cognition of the circle in ancient India, Mathematics in Higher Education (in Russian: translation by Galina Sinkevich) 14 (2016), 61–76; in: Mathematics, its Applications and History, ed. S.G. Dani, pp. 10.1–10.14, Papers contributed on the occasion of the Centenary of Ramjas College, University of Delhi; Narosa Publishing House, New Delhi, 2022.

[3] Dani, S.G., Some constructions in Mānava Śulvasūtra, Proceedings of Kanchipuram conference, in: History and Development of Mathematics

in India, pp. 373–385; Proceedings of Kanchipuram conference Ed: Sita Sundar Ram and Ramakalyani V.; published by National Mission for Manuscripts, New Delhi, co-published by D.K. Printworld (P) Ltd., New Delhi. Available at: https://arxiv.org/abs/1908.00440

[4] Dani, S.G., Mensuration of quadrilaterals in the Līlvātī, in: Bhaskaraprabha, Culture and History of Mathematics 11, Ed: K. Ramasubramanian, Takao Hayashi and Clemency Montelle, pp. 129–139 Hindustan Book Agency 2019.

[5] Datta, Bibhutibhushan, The Jaina School of Mathematics, Bull. Cal. Math. Soc. 21 (1929), 115–145.

[6] Datta, Bibhutibhushan, Mathematics of Nemicandra, The Jaina Antiquary 1 (1935).

[7] Fowler, D. and Robson, E., Square Root Approximations in Old Babylonian Mathematics: YBC 7289 in Context, Historia Math. 25 (1998) 366–378.

[8] Gupta, R.C., Bhāskara-I's approximation to sine, Indian J. Hist. Sci. 2 (1967), 121–136.

[9] Gupta, R.C., Circumference of the Jambudvīpa in Jaina Cosmography, Indian J. Hist. Sci. 10 (1973), no. 1, 38–46; (= Gaṇitānanda, pp. 93–101).

[10] Gupta, R.C., Mādhavacandra's and other octagonal derivations of the Jaina value $\pi = \sqrt{10}$, Indian J. Hist. Sci. 21 (1986), no. 2, 131–139.

[11] Gupta, R.C., On the date of *Śrīdhara*, Gaṇita Bhārati 9 (1987), 54–56; (= Gaṇitānanda, pp. 13–15).

[12] Gupta, R.C., On the volume of a sphere in ancient India, Historia Scientiarum, no. 42, 33–44 (1991); (= Gaṇitānanda, pp. 137–148).

[13] Gupta, R.C., Area of a bow-figure in India, Studies in the history of the exact sciences in honour of David Pingree, 517–532, Islam. Philos. Theol. Sci., LIV, Brill, Leiden, 2004 (= Gaṇitānanda, pp. 167–178).

[14] R.C. Gupta, Mensuration of circle according to Jaina mathematical Gaṇitānuyoga, Gaṇita Bhārati 26 (2004), no. 1–4, 131–165.

[15] Gupta, R.C., Mahāvīra-Pherū formula for the surface of a sphere and some other empirical rules, Indian J. Hist. Sci. 46 (2011), no. 4, 639–657.

[16] Heath, Thomas, A History of Greek Mathematics, Vol. II, From Aristarchus to Diophantus, Dover Publications, New York 1981.

[17] Jain, Anupam, Ardhamāgadhi sāhitya me Gaṇita (in Hindi), Jain Viśvabhāratī Viśvavidyālaya, 2008.

[18] Jain, Anupam, Mathematics in Jainism, Shrut Samvardhan Samsthan, Meerut 250002, India, 2018.

[19] Jain, Anupam, Jaina Darshan evam Gaṇita (in Hindi), Shrut Samvardhan Samsthan, Meerut 250002, India, 2019.

[20] Jain, Shefali, Mathematical contributions of Ācārya *Śrīdhara* and Ācārya *Mahāvīra*: a comparative study, Ph.D. Thesis, Shobhit Vishwavidyalaya, Meerut, India, 2013.

[21] Kichenassamy, S., Brahmagupta's derivation of the area of a cyclic quadrilateral, Historia Mathematica 37, No. 1, 2010, 28–61.

[22] Mikami, Yoshio, The Development of Mathematics in China and Japan, Chelsey Publishing House, New York (Second Edition), 1974.

[23] Neuenschwander, Erwin, Reflections on the Sources of Arabic Geometry, Sudhoffs Archive, 1988, Bd. 72, H. 2 (1988), pp. 160–169; Published by: Franz Steiner Verlag.

[24] Padmavathamma, Śrī Mahāvīracārya's *Gaṇitasārasaṅgraha*, Publ.: Śrī Siddhāntakrīthi Granthamāla, Sri Hombuja Jain Math, Hombuja, Shimoga Dist., Karnataka — 577436, India, 2000.

[25] Plofker, Kim, Mathematics in India: 500 BCE — 1800 CE, Princeton University Press, NJ, USA, 2008.

[26] Raṅgacārya, M., The *Gaṇitasārasaṅgraha* of Mahāvīracārya, published by Madras Government Press, 1912.

[27] SaKHYa (Ed), Gaṇitasārakaumudī, The Moonlight of the Essence of Mathematics by Ṭhakkura Pherū (With Introduction, Translation, and Mathemagical Commentary, Manohar Publishers, New Delhi, 2009.

[28] Saraswati Amma, T.A., Geometry in Ancient and Medieval India, Motilal Banarsidas, Delhi, Second revised edition, 1999.

[29] Sen, S.N. and Bag, A.K., The Śulvasūtras of Baudhāyana, Āpastamba, Kātyāyana, and Mānava, Indian National Science Academy, 1983.

[30] Shukla, K.S. (ed.), The Pāṭīgaṇita of Śrīdharācārya, with an ancient Sanskrit commentary, Published by Lucknow University, 1959.

[31] Shukla, K.S. and Sarma, K.V. (ed.), *Āryabhaṭīya* of Āryabhaṭa, Indian National Science Academy, New Delhi, 1976.

[32] Srikanta Sastri, S., The Date of Sridharacharya, The Jaina Antiquary, Vol. XIII, January 1948. Available at https://www.srikanta-sastri.org/the-date-of-sridharacharya/4586054224

https://doi.org/10.1142/9789811255502_0004

Article 4

Partitions of Numbers Described in *Bhagavatī Sūtra*

Dipak Jadhav

Government Boys Higher Secondary School, Anjad
Distt. Barwani (M. P.) 451556 India
dipak_jadhav17@yahoo.com

In the *Bhagavatī Sūtra*, codified for the third and last time in 466 CE, all the ways of fission possible with aggregate consisting of n ultimate-particles are enumerated when n is from two to ten, numerable, innumerable, and infinite. The enumeration involves ideas regarding partitions. The main theme of this paper is a discussion on occurrence of those ideas. It provides a detailed study of the contents of the enumeration on their own terms as well as on modern terms along with an analytical, contextual and historical point of view. On the basis of the exploration done in this paper it can be said that partitions in their true sense appeared in ancient and medieval India only in the *Bhagavatī Sūtra* unless and until any report about their appearance in any other Indian treatise comes.

Keywords: *Bhagavatī Sūtra*; compositions; partitions.

Mathematics Subject Classification 2020: 01-06, 01A32, 01A99.

4.1. Introduction

The *Bhagavaī* (Skt. *Bhagavatī* ⟨*Sūtra*⟩, "The Venerable") is one of the twelve books[1] of the Śvetāmbara Jaina canon. Its old genuine

[1]Each book (*aṇga*) has been called a *sutta* which is sanskritized as *sūtra*. The term "*aṅga*" generally refers to "addition to something previously existing". For example, there are six *Vedāṅgas* ("the ancillaries of the *Vedas*"). But the term *aṅga* in the literature of the Jainas is not "addition to something previously existing".

name is the *Viyāhapaṇṇatti*[2] (Skt. *Vyākhyāprajñapti*, "Excellent Knowledge about the Explanations" or "Suggestion on the Explanations"). *Bhagavaī* is only an epithet which was originally added to the *Viyāhapaṇṇatti* but the former has later on superseded the latter.[3] It is composed in Prakrit in prose style. It is written in the form of question and answer. It is divided into forty-one chapters. Each chapter has sub-divisions called lessons.[4]

The *Bhagavatī Sūtra* in its present form has come down to us through three stages of its development. It was first derived from the teachings of Lord Mahāvīra (599–527 BCE), the twenty-fourth and last *Tīrthaṅkara* ("Builder of a ford across an ocean of suffering") in the history of ecclesiastic Jainism, and given a shape of book by the composition of Sudharma Svāmī and then it was transmitted by him to Jambu Svāmī; both Sudharma Svāmī and Jambu Svāmī were among the eleven disciples of Lord Mahāvīra. In the second stage it was collected and fixed in the synod of monks held in 362 BCE or 358 BCE at Pāṭaliputra under the guidance of Sthūlabhadra, and in the third stage it was codified in the synod of monks held at Valabhī in 466 CE under the guidance of Devardhigaṇi Kṣamāśramaṇa.[5]

The *Bhagavatī Sūtra* is an encyclopaedic work. Along with the philosophical discourse we find brief biographies of famous contemporaries of Lord Mahāvīra, and discourses on ascetic conduct, six realities, polity, sociology, economics, history, culture, various sciences including obstetrics, mathematics, geography, and cosmology,

---

[2]At some places it is also mentioned as *Viāhapaṇṇattī* or *Viyāhapannatti* or *Vivāhapannatti*. For *Viāhapaṇṇattī* see *BhaSū*$_2$. For *Viyāhapannatti* and *Vivāhapannatti* see Deleu 1970, p. 17.

[3]Deleu 1970, p. 17.

[4]For a detailed divisions and sub-divisions of the *Bhagavatī Sūtra* see *BhaSū*$_1$, pp. xxv-xxvii. Glossary is as follows. Chapter (*saya*, Skt. *śataka*), lesson (*uddeso*, Skt. *uddeśaka*). Here the term *saya* (Skt. *śataka,* hundred) has no relevance with the number hundred. It can be said to indicate a great number of different teachings gathered in each chapter. See Schubring 1962, p. 88.

[5]Sikdar 1964, p. 34; Jain 1975, p. 33.

and miscellaneous subjects in it.[6] Those six realities include aether, anti-aether, space, soul, matter, and time.[7]

A partition of a positive integer n is a mode of expressing it as a sum of one or more positive integers, the order in which the summands occur being irrelevant. It is also called an integer partition. The summands, also called parts, are usually arranged according to size, also called order, one way or the other.[8] For example, 5 can be partitioned in seven ways:

$$5, 4+1, 3+2, 3+1+1, 2+2+1, 2+1+1+1, 1+1+1+1+1.$$

Since there is no restriction of any kind on the number or size of parts, these are called the unrestricted partitions of 5. If $p(n)$ denotes the number of unrestricted partitions of any given positive integer n, the values of $p(n)$ for the first ten positive integers are

$$p(1) = 1,\ p(2) = 2,\ p(3) = 3,\ p(4) = 5,\ p(5) = 7,$$
$$p(6) = 11,\ p(7) = 15,\ p(8) = 22,\ p(9) = 30,\ \text{and}\ p(10) = 42.$$

Partitions into odd parts, partitions into distinct parts, partitions into a specified number of parts, and partitions into a specified number of distinct parts are examples of restricted partitions. In fact, the restrictions that can be placed on the size or on the number of parts or both, are too many to enumerate.[9]

A partition is called a composition when the order of appearance of its parts matters.[10] For example, the two distinct compositions $1+3+1$ and $1+1+3$ represent the same partition $3+1+1$.

---

[6]Sikdar 1964, pp. 62–607; Pragya 2005, pp. 2–15.

[7]*BhaSū*$_2$, 13.4.55, p. 132; Sikdar 1964, p. 559. Glossary is as follows. Aether (*dhamma*, Skt. *dharma*, medium of motion), anti-aether (*adhamma*, Skt. *adharma*, medium of rest), space (*āgāsa*, Skt. *ākāśa*), soul (*jīva*), matter (*poggala*, Skt. *pudgala*), and time (*kāla*).

[8]Gupta 1970, p. 1.

[9]Gupta 1970, p. 1.

[10]It is also called a decomposition or an ordered partition.

The number of compositions is usually easier to find than that of partitions. The total number of compositions of n is given by 2^{n-1}.[11]

Partitions are also treated as the sequence of summands rather than as an expression with plus signs. For example, the partition $2 + 2 + 1$ might instead be written in tuple notation as (2, 2, 1). In order to give even more compact form this notation is further abbreviated by indicating multiplicities with exponential notation. Accordingly, (2, 2, 1) becomes $(2^2, 1)$.

An r-partition of the positive integer n is a set of r positive integers whose sum is n.[12] For example, there are just nine 4-partitions of 10. They, when expressed in compact forms, are:

$$(7, 1^3),\ (6, 2, 1^2),\ (5, 3, 1^2),\ (4^2, 1^2),\ (5, 2^2, 1),\ (4, 3, 2, 1), (3^3, 1),$$
$$(4, 2^3),\ (3^2, 2^2).$$

In the early part of the lesson four of the chapter twelve of the *Bhagavatī Sūtra* all the ways of fission possible with aggregate consisting of n ultimate-particles are enumerated when n is from two to ten, numerable (α), innumerable (β), and infinite (γ).[13] The enumeration involves ideas regarding partitions. It has not yet been studied in detail, especially from an analytical, contextual and historical point of view although it has been briefly assessed by R. S. Shah in course of his commendable and seminal survey on mathematical ideas in the *Bhagavatī Sūtra*.[14] This paper aims to

[11]Gupta 1970, p. 2.
[12]Biggs 1979, p. 124.
[13]On the basis of theorization the Jaina school of Indian mathematics is divided into the canonical class and the exclusive class. The treatises of the canonical class contain mathematics along with discussion on Jaina canons. The object of the canonical class was to demonstrate canonical thoughts including on *karma* and cosmos using mathematics. Accordingly, the *Bhagavatī Sūtra* belongs to the canonical class. For details regarding the canonical class, see Jadhav 2017, pp. 316–331. The canonical class divides positive integers into three main divisions: numerable (*saṅkhyāta*, α), innumerable (*asaṅkhyāta*, β), and infinite (*ananta*, γ). The lowest numerable number is 2 while the first innumerable number minus one is the highest numerable number. This is the number-measure system of the Jainas. For its preliminary information see Datta 1929, pp. 140–142.
[14]Shah 2008, p. 22.

study the contents of that enumeration on their own terms as well as on modern terms along with the aforesaid point of view.

4.2. Partitions as Found in the *Bhagavatī Sūtra*

According to the Jainas, ultimate-particle[15] is an indivisible part of matter. When n ultimate-particles come together, they combine to form an n-sectional aggregate. All possible partitions of an n-sectional aggregate are enumerated in the *Bhagavatī Sūtra* where n is from two to ten, numerable (α), innumerable (β) and infinite (γ). A statement is given in it for each partition. It is interesting to observe all those statements. From their study we can get to know in what style the *Bhagavatī Sūtra* presents partitions. For fear of being lengthy, here I refer to and translate the text of all the statements given for the partitions of only $n = 4$ of $n = 2, 3, \ldots, 10$, α, β, γ.

> **चउप्पएसिए खंधे भवइ। से भिज्जमाणे दुहा वि, तिहा वि, चउहा वि कज्जइ। दुहा कज्जमाणे एगयओ परमाणुपोग्गले, एगयओ तिपएसिए खंधे भवइ; अहवा दो दुपएसिया खंधा भवंति। तिहा कज्जमाणे एगयओ दो परमाणुपोग्गला, एगयओ दुपएसिए खंधे भवइ। चउहा कज्जमाणे चत्तारि परमाणुपोग्गला भवंति।**[16]

> "A tetra-sectional aggregate is formed ⟨when four ultimate-particles of matter come together and combine⟩. It divides in 2-partition or 3-partition or 4-partition if broken. It being done in 2-partition, on one side becomes a ⟨single⟩ ultimate-particle of matter and on the other side becomes a tri-sectional aggregate; or there become two ⟨separate⟩ bi-sectional aggregates. It being done in 3-partition, on one side are two ⟨separate⟩ ultimate-particles of matter and on the other side is a ⟨single⟩ bi-sectional aggregate. It being

[15]The English term "atom" cannot and must not be taken to be equivalent to ultimate-particle (*parama-aṇu* or *paramāṇu*). It is defined in modern science to consist of a positively charged nucleus consisting of protons and neutrons surrounded by a cloud of electrons charged negatively. However, its Greek root is *atomos*, which means "indivisible". The scientists who first gave the atom its name imagined it could not be split into smaller pieces.

[16]*BhaSū*$_5$, 12.4.4, p. 267.

done in 4-partition, all the four ultimate-particles of matter get separated."[17]

On the basis of the above I deduce that (i) $p_{BhaS\bar{u}}$ $(4) = 4$ where $p_{BhaS\bar{u}}$ (n) is the total number of partitions enumerated in the *Bhagavatī Sūtra* for n and (ii) partitions are expressed in it as the sequence of summands, that too in compact form, rather than as an expression with plus signs. Therefore, those partitions that are enumerated in the *Bhagavatī Sūtra* for the value of n from 2 to 10 can be expressed as follows.[18]

---

[17]Glossary is as follows. Tetra-sectional aggregate (*caˆuppaˆesie khaṃdha*, Skt. *catuṣpradeśika skandha*), 2-partition (*duhā*, Skt. *dvidhā*, twofold), 3-partition (*tihā*, Skt. *tridhā*, threefold), 4-partition (*caˆuhā*, Skt. *caturdhā*, fourfold), ultimate-particle of matter (*paramāṇupoggala*, Skt. *paramāṇupudgala*), tri-sectional aggregate (*tipaesie khaṃdha*, Skt. *tripradeśika skandha*), bi-sectional aggregate (*dupaesiye khaṃdha*, Skt. *dvipradeśika skandha*). Amar Muniji *et al.* interpreted the term *paesa* (Skt. *pradeśa*), which originally means space-point, appearing in *dupaesiye* (Skt. *dvipradeśika*), *tipaesie* (Skt. *tripradeśika*), and *caˆuppaˆesie* (*catuṣpradeśika*), as section while Zaveri and Kumar, following the Greek sense of atom for ultimate-particle, as atom. According to the interpretation offered by Amar Muniji *et al.* "*dupaesiye khaṃdha*" means "bi-sectional aggregate" while it means "di-atomic aggregate" according to the one offered by Zaveri and Kumar. See *BhaSū*$_5$, p. 268; Zaveri and Kumar 1995, p. 105. I have followed Amar Muniji *et al.* here. It does not mean that the interpretation offered by Zaveri and Kumar is not appropriate. In fact, the term *paesa* is merely conceptual here. The explanation of why it is conceptual is as follows. *Astikāya* is a compound term made up of *asti* (existence) and *kāya* (extensive body). It thus means a real extensive magnitude. It has plurality of sections (*paesas*, Skt. *pradeśas*) in its constitution. Similarly, *pudgala* (matter) is a compound term made up of *pud* (to integrate) and *gala* (to disintegrate). It thus is that which undergoes modifications by integration and disintegration. *Pudgala* is an *astikāya* as its all aggregates are composed of multiple sections (*paesas*). Its extension in space varies from aggregate to aggregate depending upon its density (i.e., sections). Ultimate-particles are the ultimate building blocks which by mutual combination produce the whole of physical universe. Ultimate-particle is a space-point so long as it is considered to be a section of an aggregate while it is an ultimate-particle in its free state. See Zaveri and Kumar 1995, pp. 75, 85, 95, and 114.

[18]*BhaSū*$_5$, 12.4.2-10, pp. 266–284.

When $n = 2$, 2-partition: (1^2).
Therefore, $p_{BhaS\bar{u}}$ $(2) = 1$.
When $n = 3$, 2-partitions: (1, 2);
3-partition: (1^3).
Therefore, $p_{BhaS\bar{u}}$ $(3) = 2$.
When $n = 4$, 2-partitions: (1, 3), (2, 2);
3-partition: $(1^2, 2)$;
4-partition: (1^4).
Therefore, $p_{BhaS\bar{u}}$ $(4) = 4$.
When $n = 5$, 2-partitions: (1, 4), (2, 3);
3-partitions: $(1^2, 3)$, $(1, 2^2)$;
4-partition: $(1^3, 2)$;
5-partition: (1^5).
Therefore, $p_{BhaS\bar{u}}$ $(5) = 6$.
When $n = 6$, 2-partitions: (1, 5), (2, 4), (3, 3);
3-partitions: $(1^2, 4)$, (1, 2, 3), (2, 2, 2);
4-partition: $(1^3, 3)$, $(1^2, 2^2)$;
5-partition: $(1^4, 2)$;
6-partition: (1^6).
Therefore, $p_{BhaS\bar{u}}$ $(6) = 10$.
When $n = 7$, 2-partitions: (1, 6), (2, 5), (3, 4);
3-partitions: $(1^2, 5)$, (1, 2, 4), (1, 3, 3), (2, 2, 3);
4-partitions: $(1^3, 4)$, $(1^2, 2, 3)$, $(1, 2^3)$;
5-partitions: $(1^4, 3)$, $(1^3, 2^2)$;
6-partition: $(1^5, 2)$;
7-partition: (1^7).
Therefore, $p_{BhaS\bar{u}}$ $(7) = 14$.
When $n = 8$, 2-partitions: (1, 7), (2, 6), (3, 5), (4, 4);
3-partitions: $(1^2, 6)$, (1, 2, 5), (1, 3, 4), (2, 2, 4), (2, 3, 3);
4-partitions: $(1^3, 5)$, $(1^2, 2, 4)$, $(1^2, 3^2)$, $(1, 2^2, 3)$, (2^4);
5-partitions: $(1^4, 4)$, $(1^3, 2, 3)$, $(1^2, 2^3)$;
6-partitions: $(1^5, 3)$, $(1^4, 2^2)$;
7-partition: $(1^6, 2)$;

8-partition: (1^8).
Therefore, $p_{BhaS\bar{u}}$ $(8) = 21$.

When $n = 9$, 2-partitions: (1, 8), (2, 7), (3, 6), (4, 5);
3-partitions: $(1^2, 7)$, (1, 2, 6), (1, 3, 5), (1, 4, 4), $(2^2, 5)$, (2, 3, 4), (3, 3, 3);
4-partitions: $(1^3, 6)$, $(1^2, 2, 5)$, $(1^2, 3, 4)$, $(1, 2^2, 4)$, $(1, 2, 3^2)$, $(2^3, 3)$;
5-partitions: $(1^4, 5)$, $(1^3, 2, 4)$, $(1^3, 3^2)$, $(1^2, 2^2, 3)$, $(1, 2^4)$;
6-partitions: $(1^5, 4)$, $(1^4, 2, 3)$, $(1^3, 2^3)$;
7-partitions: $(1^6, 3)$, $(1^5, 2^2)$;
8-partition: $(1^7, 2)$;
9-partition: (1^9).
Therefore, $p_{BhaS\bar{u}}$ $(9) = 29$.

When $n = 10$, 2-partitions: (1, 9), (2, 8), (3, 7), (4, 6), (5, 5);
3-partitions: $(1^2, 8)$, (1, 2, 7), (1, 3, 6), (1, 4, 5), $(2^2, 6)$, (2, 3, 5), (2, 4, 4), (3, 3, 4);
4-partitions: $(1^3, 7)$, $(1^2, 2, 6)$, $(1^2, 3, 5)$, $(1^2, 4^2)$, $(1, 2^2, 5)$, (1, 2, 3, 4), $(1, 3^3)$, $(2^3, 4)$, $(2^2, 3^2)$;
5-partitions: $(1^4, 6)$, $(1^3, 2, 5)$, $(1^3, 3, 4)$, $(1^2, 2^2, 4)$, $(1^2, 2, 3^2)$, $(1, 2^3, 3)$, (2^5);
6-partitions: $(1^5, 5)$, $(1^4, 2, 4)$, $(1^4, 3^2)$, $(1^3, 2^2, 3)$, $(1^2, 2^4)$;
7-partitions: $(1^6, 4)$, $(1^5, 2, 3)$, $(1^4, 2^3)$;
8-partition: $(1^7, 3)$, $(1^6, 2^2)$;
9-partition: $(1^8, 2)$;
10-partition: (1^{10}).
Therefore, $p_{BhaS\bar{u}}$ $(10) = 41$.

4.3. Results and Discussion

Comparing the corresponding values of $p(n)$ and $p_{BhaS\bar{u}}$ (n) when (n) is from 2 to 10, we have

$$p(n) = p_{BhaS\bar{u}}(n) + 1.$$

It was also noticed by R. S. Shah.[19] But he did not explain why the *Bhagavatī Sūtra* enumerated one partition less. The reason for not offering 1-partition in each case relates to the context in which ideas regarding partitions were developed in it. Ultimate-particles form an aggregate when they come together and combine. The aggregate, as we are able to see above, may divide in at least two parts and at most in as many parts as there are ultimate-particles combined in it. Those parts are separate ultimate-particles or separate aggregates containing "a number of ultimate-particles smaller than that of the original aggregate" or both.[20] Since importance is given in the *Bhagavatī Sūtra* to the division of an aggregate, the aggregate is not considered to be a division of itself. That is why 1-partition was not offered in the *Bhagavatī Sūtra* in any case. On the other hand, the number 'n' itself is regarded in modern combinatorics and number theory as one partition for the sum is the summand itself if the summation has one summand.[21] The sum is zero if the summation has no summand, because zero is the additive identity. In other words, an empty sum is a summation where the number of summands is zero.[22]

We can also interpret $p(n) = p_{BhaS\bar{u}}(n) + 1$ in modern combinatorial terms as follows. The partitions enumerated in the *Bhagavatī*

---

[19]Shah 2008, p. 22.

[20]Jozef Deleu [1970, p. 183] also noticed this.

[21]The number 'n' seems to have not been treated as one partition even in the very early stages of modern thinking of partitions as it is evident from the following. "In a letter to Johann Bernoulli, dated July 28, 1699, G. W. Leibniz asked if he had investigated the number of ways in which a given number could be broken up into two, three, etc., pieces. It seems that this is not an easy problem, and yet it would be worth knowing." See Gupta 1970, p. 1; Biggs 1979, pp. 126–127. From this curiosity, theory of partitions seems to have sprouted in Europe. Later, Euler heightened the ideas sketched by Leibniz to a full-grown theory of partitions. A serious study of the subject of partitions started with him in modern time. It is noticeable that the number of ways about which Leibniz asked Bernoulli does not include the way in which a given number can be broken up into one piece or itself.

[22]Harper 2016, p. 86.

Sūtra are the restricted ones and the restriction imposed therein is $2 \leq r \leq n$.

Another noticeable point is that no partition is referred to in the *Bhagavatī Sūtra* for $n = 1$. It also relates to the context. If attention was given in the *Bhagavatī Sūtra* to finding $p_{BhaS\bar{u}}$ (1), it would be contrary to the attributes of ultimate-particle. It is described in the *Bhagavatī Sūtra* that ultimate-particle is uncleavable, impenetrable, incombustible, intangible, without half, without inner part, without space-point, and partitionless.[23]

It is conventional in modern combinatorics to write the parts of a partition in descending order although the order is irrelevant. The convention of writing the parts of a partition in the *Bhagavatī Sūtra* being just opposite to what is in modern combinatorics is noticeable. Perhaps the reason for adopting the ascending order in the *Bhagavatī Sūtra* is that priority is given in it to finding out the number of "separate ultimate-particles" over that of "separate aggregates containing a number of ultimate-particles smaller than that of the original aggregate" in which the original aggregate divides.

We are able to see that two partitions are marked in section two of this paper by drawing a line underneath each of them. The first is (2^2, 5) in 3-partitions[24] of 9 and the second is (1, 2^2, 5) in 4-partitions[25] of 10. Both of them are missing in the *Bhagavatī Sūtra.* R. S. Shah paid attention to the latter only. He argues that this may be due to mistake of copyists of manuscripts of later years.[26] Since the method put into practice in the *Bhagavatī Sūtra* for enumerating partitions of from 2 to 8 is precise and is extendable to enumerate

---

[23] *BhaSū*$_4$, 20.5.29-30, p. 561; Sikdar 1964, p. 568; Deleu 1970, p. 113. Glossary is as follows. Uncleavable (*acchejja*, Skt. *achedya*), impenetrable (*abhejja*, Skt. *abhedya*), incombustible (*aḍajjha*, Skt. *adāhya*), intangible (*agejjha*, Skt. *agrāhya*), without half (*aṇaddha*, Skt. *anarddha*, no halves), without inner part (*amajjha*, Skt. *amadhya*, no middle), without space-point (*apaesa*, Skt. *apradeśa*, no unit of space, i.e., dimensionless), and partitionless (*avibhāima*, Skt. *avibhājya*, indivisible).

[24] *BhaSū*$_5$, 12.4.9, p. 276; *BhaSū*$_2$, 12.4.76, pp. 33–34.

[25] *BhaSū*$_5$, 12.4.10, p. 280; *BhaSū*$_2$, 12.4.77, p. 36.

[26] Shah 2008, p. 22.

partitions of an integer greater than 8, I endorse his argument for both of the missing partitions. It is interesting to note that Sādhvī Līlamabāī has inserted those two partitions by writing them in Prakrit at proper places in her edited version of the *Bhagavatī Sūtra* and has marked her insertions by placing them in square brackets.[27]

We have seen in the previous section that the terms employed for 2-partition, 3-partition, and 4-partition are *duhā*, *tihā*, and *caˆuhā* respectively.[28] It is intelligible that the pattern followed for naming r-partition is that *hā*[29] is suffixed to r. The terms used in the *Bhagavatī Sūtra* from 5-partition to 10-partition are *paṃcahā*, *cahā*, *sattahā*, *aṭṭhahā*, *navahā*, *dasahā* respectively.[30] Terms like numerable-partition (α-partition), innumerable-partition (β-partition), and infinite-partition (γ-partition) are also used in the *Bhagavatī Sūtra.*[31]

We succeeded to identify the pattern given in the *Bhagavatī Sūtra* for naming r-partition through the above discussion. But we could not find any general term in it for partition from the description given in section two of this paper. The terms used as a general term for partition by the editors Ācārya Mahāprajña[32], Amar Muniji Maharaj *et al.*[33] and Līlamabāī Mahāsatī *et al.*[34] in their annotations and notes are *bheda*, *vikalpa*, and both *vikalpa* and *bhaṅga* respectively.

---

[27] For (2^2, 5) see *BhaSū*$_3$, 12.4.8, p. 677. For (1, 2^2, 5) see *BhaSū*$_3$, 12.4.16, p. 681.

[28] Glossary is as follows. *Duhā* (Skt. *dvidhā*, twofold), *tihā* (Skt. *tridhā*, threefold), and *caˆuhā* (Skt. *caturdhā*, fourfold).

[29] The Sanskrit equivalent term for *hā* is *dhā* (fold). See Apte 1893, p. 321.

[30] *BhaSū*$_5$, 12.4.10, pp. 280–281. Glossary is as follows. *paṃcahā* (Skt. *pañcadhā*, fivefold), *cahā* (Skt. *ṣaḍḍhā*, sixfold), *sattahā* (Skt. *saptadhā*, sevenfold), *aṭṭhahā* (Skt. *aṣṭadhā*, eightfold), *navahā* (Skt. *navadhā*, ninefold), and *dasahā* (Skt. *daśadhā*, tenfold).

[31] *BhaSū*$_5$, 12.4.11-13, pp. 286–294. Glossary is as follows. Numerable-partition (*saṃkhejjahā*, Skt. *saṅkhyeyadhā*, numerable-fold), innumerable-partition (*asaṃkhejjahā*, Skt. *asaṅkhyeyadhā*, innumerable-fold), and infinite-partition (*aṇaṃtahā*, Skt. *anantadhā*, infinite-fold).

[32] *BhaSū*$_2$, pp. 42–44.

[33] *BhaSū*$_5$, pp. 284–285.

[34] *BhaSū*$_3$, p. 684.

As far as *vikalpa* and *bhaṅga* are concerned, B. B. Datta finds them to be the ancient Indian terms for permutations and combinations respectively.[35] The term *bheda* seems to have been conceived as a general term in the *Bhagavatī Sūtra* for naming partition as it is associated with *bhijjamāṇe* that comes up frequently in the texts, including the one referred to for $n = 4$ in the previous section of this paper, pertaining to the enumeration of partitions of from 2 to γ.[36] What is in support of this argument of mine is that Bhāskara II (born in 1114 CE and died after 1183 CE) also used the term *vibheda* for composition. Still, I am not in a position to say that neither of *vikalpa* and *bhaṅga* can be used as a general term for partition. In fact, calling the combinatorial concept by a general term appears to be an issue in the study of ancient Indian combinatorics, which needs to be addressed.[37]

The last chapter of the *Līlāvatī* ("Gorgeous Amusement ⟨of Mathematics⟩") of Bhāskara II is on concatenation. One of its rules is relevant here. The rule is as follows.

निरेकमंकैक्यमिदं निरेकस्थानांतमेकापचितं विभक्तम्।
रूपादिभिस्तन्निहतेः समा स्युः संख्याविभेदा नियतेऽङ्कयोगे॥
नवान्वितस्थानकसंख्यकायाः ऊनेऽङ्कयोगे कथिते तु वेद्यम्।
संक्षिप्तमुक्तं पृथुताभयेन नान्तोऽस्ति यस्माद् गणितार्णवस्य॥.[38]

"When the sum (n) of the digits is fixed, divide the numbers beginning with the sum minus one and lessening by one, that are set down in a row in as many places (r) less one as given, by one, etc. respectively and multiply those quotients together. The product

[35]Datta and Singh 1992, p. 232. Glossary is as follows. *Vikalpa* (alternatives), and *bhaṅga* (poses).

[36]Glossary is as follows. *bheja* (Skt. *bheda*, division or separation), and *bhijjamāṇe* (Skt. *bhidyamānaḥ*).

[37]I propose to write a separate paper to address such an issue, which frequently comes up in the study of ancient Indian combinatorics.

[38]*Lī*, vv. 274–275, p. 180. Many scholars recognize that the rule is on partitions. See *Lī*, p. 181; Wagner 2018, p. 107. But it does not pertain to partitions in the sense in which partitions are defined.

will be the number of ⟨r-⟩ compositions ⟨of the sum⟩. This rule is applicable only when the sum (n) of those digits is less than the number of the places (r) plus nine. For fear of being long this is treated in brief for ocean of mathematics is without bounds."[39]

That is, the total number of r-compositions of a positive integer n is given by

$$\frac{(n-1)}{1} \times \frac{(n-2)}{2} \times \cdots \times \frac{(n-(r-1))}{(r-1)} \quad \text{or} \quad \frac{(n-1)!}{(n-r)!(r-1)!}.$$

Bhāskara II further gives an example to find 5-compositions of 13, which comes to be 495 when calculated.[40]

Nārāyaṇa Paṇḍita wrote his *Gaṇitakaumudī* ("Moon Light of Mathematics") in 1356 CE. In its chapter thirteen, he gives a number of rules accompanied by examples on r-compositions of n. Those rules are for finding out the total number of r-compositions, the total number of r-compositions ending in a particular digit, the number of a particular digit and number of all digits in r-compositions, the sum of r-compositions, the sum of all compositions, the structures of an r-composition, the r-composition when its serial number is given, and the serial number of an r-composition when its structure is given.[41]

It may be of interest here to recall that the *Bhagavatī Sūtra* discusses also enumerations of the partitions of α, β, and γ in an indicative and suggestive manner by taking r from 2 to 11, 12, and 13 respectively. Those enumerations are as follows.[42]

---

[39]Glossary is as follows. Sum (*yoga*), digits (*aṅkas*), fixed (*niyata*), places (*sthānas*), number (*saṅkhyā*), ⟨r-⟩ compositions (*vibhedas*), places (*sthānakas*), ocean (*arṇava*), and mathematics (*gaṇita*).

[40]*Lī*, v. 276, p. 181.

[41]Singh 2001, pp. 62–73. Parmanand Singh writes that "partition dealt with by Bhāskara II and Nārāyaṇa slightly differs from partition used in modern mathematics." See Singh 2001, p. 62. This statement of Singh shows that the rules given by Bhāskara II and Nārāyaṇa are not related to partitions. In fact, they are related to compositions.

[42]*BhaSū*$_5$, 12.4.11-13, pp. 285–296; *BhaSū*$_3$, 12.4.22-33, pp. 685–693.

When n is α, 2-partitions: $(1, \alpha)$, $(2, \alpha)$, $(3, \alpha)$, $\ldots$, $(10, \alpha)$, (α, α); total $= 11$;
3-partitions: $(1^2, \alpha)$, $(1, 2, \alpha)$, $(1, 3, \alpha)$, $\ldots$, $(1, 10, \alpha)$, $(1, \alpha^2)$, $(2, \alpha^2)$, $\ldots$, $(10, \alpha^2)$, (α^3); total $= 21$;
4-partitions: $(1^3, \alpha)$, $(1^2, 2, \alpha)$, $(1^2, 3, \alpha)$, $\ldots$, $(1^2, 10, \alpha)$, $(1^2, \alpha^2)$, $(1, 2, \alpha^2)$, $\ldots$, $(1, 10, \alpha^2)$, $(1, \alpha^3)$, $\ldots$, $(10, \alpha^3)$, (α^4); total $= 31$; similarly,
5-partitions: total $= 41$;
6-partitions: total $= 51$;
7-partitions: total $= 61$;
8-partitions: total $= 71$;
9-partitions: total $= 81$;
10-partitions: $(1^9, \alpha)$; $(1^8, 2, \alpha)$, $\ldots$, $(10, \alpha^9)$; (α^{10}); total $= 91$; α-partition: (1^α); total $= 1$.
Therefore, $p_{BhaS\bar{u}}(\alpha) = 460$.

When n is β, 2-partitions: $(1, \beta)$, .., $(10, \beta)$, (α, β), (β^2); total $= 12$;
3-partitions: $(1^2, \beta)$, $(1, 2, \beta)$, $\ldots$, $(1, 10, \beta)$, $(1, \alpha, \beta)$, $(1, \beta^2)$, $(2, \beta^2)$, $\ldots$, (α, β^2), (β^3); total $= 23$;
4-partitions: $(1^3, \beta)$, $\ldots$; total $= 34$;
5-partitions: total $= 45$;
6-partitions: total $= 56$;
7-partitions: total $= 67$;
8-partitions: total $= 78$;
9-partitions: total $= 89$;
10-partitions: total $= 100$;
α-partition: $(1^\alpha, \beta)$, $(2^\alpha, \beta)$, $\ldots$, $(10^\alpha, \beta)$, (α^α, β), (β^α); total $= 12$;
β-partition: (1^β); total $= 1$.
Therefore, $p_{BhaS\bar{u}}(\beta) = 517$.

When n is γ, 2-partitions: $(1, \gamma), \ldots, \langle(10, \gamma), (\alpha, \gamma), (\beta, \gamma),\rangle\ (\gamma^2)$; total = 13;
3-partitions: $(1^2, \gamma), (1, 2, \gamma), \ldots, (1, \beta, \gamma), (1, \gamma^2), (2, \gamma^2), \ldots, (10, \gamma^2), (\alpha, \gamma^2), (\beta, \gamma^2), (\gamma^3)$; total = 25;
4-partitions: $(1^3, \gamma), \ldots$; total = 37;
5-partitions: total = 49;
6-partitions: total = 61;
7-partitions: total = 73;
8-partitions: total = 85;
9-partitions: total = 97;
10-partitions: total = 109;
α-partition: $(1^\alpha, \gamma), (2^\alpha, \gamma), \ldots, (10^\alpha, \gamma), (\alpha^\alpha, \gamma), (\beta^\alpha, \gamma), (\gamma^\alpha)$; total = 13;
β-partition: $(1^\beta, \gamma), (2^\beta, \gamma), \ldots, (\alpha^\beta, \gamma), (\beta^\beta, \gamma), (\gamma^\beta)$; total = 13;
γ-partition: (1^γ); total = 1.
Therefore, $p_{BhaS\bar{u}}(\gamma) = 576$.

Before concluding, I would like to record that the topic of partitions has been much studied since the last century. In particular, in 1918 CE, Hardy and Ramanujan proved that $p(n) \approx \frac{e^c}{4n\sqrt{3}}$ where $c = \pi\sqrt{\frac{2n}{3}}$, which gives general order of magnitude of $p(n)$ as $n \to \infty$.[43] An exact formula for finding $p(n)$ was obtained in 1937 CE by Hans Rademacher. $p(n)$ is the coefficient of x^n in the expansion of $[(1-x)(1-x^2)(1-x^3)\ldots]^{-1}$.[44]

4.4. Concluding Remarks

Partitions found in the *Bhagavatī Sūtra* are in applied form. Ideas regarding them must have been developed in it on the basis of actual additive separation, that too keeping their application in mind. They are found to be faithful to the context under and for which partitions

[43]Debnath 1987, p. 632.
[44]Gupta 1970, p. 14; Debnath 1987, p. 633.

of n from 2 to 10 and in a suggestive manner from α to β and γ are actually enumerated in it. Compositions are involved in the *Līlāvatī* of Bhāskara II and the *Gaṇitakaumudī* of Nārāyaṇa Paṇḍita in certain combinatorial problems. But there is no enumeration of partitions in them. Therefore, it can be said that in ancient and medieval India partitions are found only in the *Bhagavatī Sūtra* unless and until any report about their appearance in any other Indian treatise comes.

Abbreviation and Notation

Skt. Sanskrit. The term(s) put by me just after Skt. will help the reader to understand Prakrit through Sanskrit.

⟨...⟩ A pair of pointing angle quotation marks, wherever used, contains a paraphrase supplied by me to achieve comprehensiveness together with clarity. It does not mean that the original expressions are incomplete or corrupted.

Acknowledgments

Except for a few changes including its title, this paper was presented in 27th International Conference of International Academy of Physical Sciences on *Mathematics and Science in Ancient India* organized by Vikram University (Ujjain) in association with International Academy of Physical Sciences (Prayagraj) during October 26–28, 2021 (Online Mode). I would like to place on record my thanks to the organizers for giving me the opportunity to present the paper. I wish to thank the referee for the valuable comments and suggestions. I am highly grateful to Professor Surender Kumar Jain (Ohio) for making this paper a part of this monograph.

References

Primary sources

[1] *BhaSū*$_1$ *Bhagavaī*, Vol. I (Śataka I to Śataka II), Edited and annotated by Ācārya Mahāprajña and translated into

English by Muni Mahendra Kumar (Ladnun: Jain Vishva Bharati, 2005).

[2] *BhaSū*$_2$ *Bhagavaī*, Vol. IV (Śataka 12 to Śataka 16), Edited and annotated by Ācārya Mahāprajña (Ladnun: Jain Vishva Bharati, 2007).

[3] *BhaSū*$_3$ *Śrī Bhagavatī Sūtra,* Vol. III (Śataka 8 to Śataka 12), Edited by Līlamabāī Mahāsatī with the assistance of Sādhvī Āratībāī and Sādhvī Subodhikābāī and translated into Gujrati by Sādhvī Āratībāī (Mumbai: Śrī Guru Prāṇa Prakāśana, 2009).

[4] *BhaSū*$_4$ *Śrī Bhagavatī Sūtra,* Vol. IV (Śataka 13 to Śataka 23), Edited by Līlamabāī Mahāsatī with the assistance of Sādhvī Āratībāī and Sādhvī Subodhikābāī and translated into Gujrati by Sādhvī Āratībāī (Mumbai: Śrī Guru Prāṇa Prakāśana, 2009).

[5] *BhaSū*$_5$ *Illustrated Shri Bhagavati Sutra (Vyakhya Prajnapti)* (Original Text with Hindi and English Translations, Elaboration and Multicoloured Illustrations), Vol. IV (Chapters 10-13), Edited by Amar Muniji Maharaj, Varun Muni, and Sanjay Surana, translated into English by S. Bothara, and illustrated by T. Sharma (Delhi: Padma Prakāśana, 2013).

[6] *Lī* *Līlāvatī of Bhāskarācārya*, A treatise of mathematics of Vedic tradition with rationales in terms of modern mathematics largely based on N. H. Phadke's Marāṭhī translation of *Līlāvatī*, Translated by Krishnaji Shankar Patwardhan, Somashekhara Amrita Naimpally and Shyam Lal Singh into English (Delhi: Motilal Banarsidass, 2001).

Secondary literature

[7] Vaman Shivram Apte, *The Student's English-Sanskrit Dictionary* (Bombay: Mrs. Radhabai Atmaram Sagoon, 1893).

[8] N. L. Biggs, The roots of combinatorics, *Historia Mathematica* 6 (1979) 109–136.

[9] B. B. Datta, The Jaina school of mathematics, *Bulletin of the Calcutta Mathematical Society* 21 (1929) 115–145.

[10] B. B. Datta and A. N. Singh, Use of permutations and combinations in India (Revised by Kripa Shankar Shukla), *Indian Journal of History of Science* 27.3 (1992) 231–249.

[11] Lokenath Debnath, Srinivasa Ramanujan (1887-1920) and the theory of partitions of numbers and statistical mechanics: A centennial tribute, *International Journal of Mathematics and Mathematical Sciences* 10.4 (1987) 625–640.

[12] Jozef Deleu, *Viyāyapannatti (Bhagavaī); The Fifth Anga of the Jaina Canon : Introduction, Critical Analysis, Commentary & Indexes* (Brugge: De Tempel, 1970).

[13] Hansraj Gupta, Partitions: A survey, *Journal of Research of the National Bureau of Standards: B. Mathematical Sciences* 74B-1 (1970) 1–29.

[14] Robert Harper, *Practical Foundations for Programming Languages* (Cambridge University Press, 2016).

[15] Dipak Jadhav, The Jaina school of Indian mathematics, *Indian Journal of History of Science* 52.3 (2017) 316–334.

[16] Subodh Kumar Jain, Chronology of ancient Jaina conferences, *The Jaina Antiquary* 27.1 (1975) 31–34.

[17] Samani Chaitanya Pragya, *Scientific Vision of Lord Mahāvīra* (Ladnun: Jain Vishva Bharati, 2005).

[18] Walther Schubring, *The Doctrine of the Jainas* (*Described after the Old Sources*) Translated from the revised German edition by Wolfgang Beurlen (Delhi: Motilal Banarsidass, 1962). The original edition in German was published in 1934 under the title *Die Lehre der Jainas, nach den alten Quellen dargestellt.*

[19] R. S. Shah, Mathematical ideas in Bhagavatī Sūtra, *Gaṇita Bhāratī* 30.1 (2008) 1–25.

[20] Jogendra Chandra Sikdar, *Studies in the Bhagavatī Sūtra* (Muzaffarpur: Research Institute of Prakrit, Jainalogy & Ahımsa, 1964).

[21] Parmanand Singh, The Gaṇita Kaumudī of NārāyaṇaPaṇḍita: Chapter XIII (English translations with notes), *Gaṇita Bhāratī* 23(1-4) (2001) 18-82.

[22] Roy Wagner, The *Kriyākramakarī's* integrative approach to mathematical knowledge, *History of Science in South Asia* 6 (2018) 84–126.

[23] Jethalal S. Zaveri and Muni Mahendra Kumar, *Microcosmology: Atom in The Jain Philosophy and Modern Science* (Ladnun: Jain Vishva Bharati Institute, 1995).

https://doi.org/10.1142/9789811255502_0005

Article 5

Concept of Infinity in Jaina Literature

Jitendra K. Sharma

Department of Mathematics
Shri Rawatpura Sarkar University, Raipur
sharma_jit2000@yahoo.com

The number system of twenty-one folds and the eleven types of infinity referred to in Jaina literature are briefly described in this paper.

Keywords: Dhavalā, infinity, number system of twenty-one folds

Mathematics Subject Classification 2020: 01-06,01A32, 01A99

5.1. Introduction

It is universally accepted that the concept of infinity is the original contribution of ancient Indian mathematicians [1]. There is another aspect to the idea of infinity, namely, counting infinity. In modern mathematics, this leads to the concepts of ordinal and cardinal numbers.

In 628 CE, Brahmagupta in his *Brāhmasphuṭasiddhānta* gave rules for arithmetic operations involving zero. He explained the mathematical operations like addition, subtraction and multiplication involving zero but when it came to division by zero, he gave some rules that were not correct [2]. According to him, a zero divided by a zero is zero; a negative or a positive number divided by zero has that (zero) as its divisor, or zero divided by a negative or a positive has that negative or positive number as its divisor.

Note that a non-zero quantity divided by zero seems to remain somehow "zero-divided"; it is not clear whether the quantity is considered unchanged by the division by zero [3]. After 500 years of Brahmagupta, Bhāskarācārya discussed the problem of division, stating that any number divided by zero is infinity. In the rules for the arithmetic of zero described in his work Bījagaṇita, there is an explicit association of division by zero with infinity, which was absent in the Līlāvatī [4]. In 850 CE, Mahāvīrācārya wrote *Gaṇitasārasaṅgraha*, He correctly stated the multiplication rules for zero but fails to explain properly division by zero [5]. In ancient Indian mathematics, we find *Anuyogadvāra sūtra* (3rd century CE) and the text written there after discussing various such concepts of infinities. These texts are mainly religious or philosophical, but often carry a healthy amount of serious mathematics. They seem to introduce formal concepts of numerable, innumerable and infinite. They even classify multidimensional concepts for infinity. It is possible that they might have come close to the ideas of modern cardinal (or at least ordinal) numbers. A.N. Singh pointed out about one-to-one correspondence which form the basis of the modern theory of infinite cardinals [6].

In the present paper efforts are made to describe the classification of numbers and infinity in Jaina texts. We find a great fascination for very large numbers among Jaina ascetics. Their cosmology, cosmogony and *Karma* theory involved dealing with very large numbers. In the present article the eleven types of infinity as found in *Dhavalā* of Vīrasena (816 CE) will be discussed.

Ācārya Vīrasena (816 CE) [7] was a Digambara Jaina ascetic and belonged to the lineage of Ācārya Kundakunda [8]. His most reputed work is *Dhavalā*. Most of the mathematical content of *Dhavalā* belongs to the period 200 to 600 CE [9]. Therefore, this text is most important to historians of Indian mathematics, as it delivers the information about the period earlier to 600 CE. The author of *Dhavalā* is fully conversant with the place value system of notation [10], fundamental operations, extraction of square and cube roots [11], raising of a number to its own power [12], laws of indices [13], logarithms [14]. Vīrasena also explains large numbers and the methods of expressing them and classification of infinites efficiently.

5.2. Classification of Numbers

In Jaina texts, an enumeration goes to *Śirṣaprahelikā*, a number with 194 digits [15]. In *Anuyogadvārasūtra* and *Bhagavatīsūtra* there is a significant remark that arithmetic ends here and further enumeration can be understood through simile measures viz. *palyopama* [16] and *sāgaropama* [17]. They classified positive integers excluding one into three groups [18] — *saṅkhyāta* (numerable), *asaṅkhyāta* (innumerable), and *ananta* (infinite). Each of them is subdivided into three orders:

1. Numerable: *jaghanya* (lowest), *madhyama* (intermediate), *utkṛṣṭa* (highest).
2. Innumerable: *paritāsāṅkhyāta* (peripheral innumerable), *yuktāsāṅkhyāta* (yoked innumerable), and *asaṅkhyātāsaṅkhyāta* (innumerably innumerable).
3. Infinite: *paritānanta* (peripheral infinity), *yuktānanta* (yoked infinity), and *anantānanta* (infinitely infinite).

Moreover, each of the divisions of *asaṅkhyāta* and *ananta* is further divided into 3 sub-categories *jaghanya, madhyama, utkṛṣṭa* [19].

Thus there are 21 classes of numbers. The *jaghanya saṅkhyāta* is the lowest number while the *utkṛṣṭa anantānanta* is the highest. For details see Table [21].

The *jaghanya saṅkhyāta* is 2. While describing numerable *Anuyogadvāra sūtra* starts from number 2. 1 was not regarded as number for counting purposes [22]. The madhyama saṅkhyāta includes all numbers between 2 and the *utkṛṣṭa saṅkhyāta* which itself is a number immediately preceding the *jaghanya paritāsaṅkhyāta*. In *Trilokasāra* [23], the *jaghanya parita asaṅkhyāta* is explained in the following manner.

Table 5.1. Jaina classification of numbers

S. No.	Number		Interpretation
1	Jaghanya saṅkhyāta	Lowest numerable	2
2	Madhyama saṅkhyāta	Intermediate numerable	$3, 4, \ldots, \alpha - 2$
3	Utkṛṣṭa saṅkhyāta	Highest numerable	$\alpha - 1$
4	Jaghanya *paritasaṅkhyāta*	Lowest lower order innumerable	α (as defined)
5	Madhyama *paritasaṅkhyāta*	Intermediate middle order innumerable	$\alpha + 1, \ldots, \alpha^{\alpha}$-2
6	utkṛṣṭa *paritasaṅkhyāta*	Highest higher order innumerable	α^{α}-1
7	Jaghanya *yuktasaṅkhyāta*	Lowest middle order innumerable	α^{α}
8	Madhyama *yuktasaṅkhyāta*	Intermediate middle order innumerable	$\alpha^{\alpha} + 1, \ldots, \alpha^{2\alpha}$-2
9	utkṛṣṭa *yuktasaṅkhyāta*	Highest middle order innumerable	$\alpha^{2\alpha}$-1
10	Jaghnaya *asaṅkhyātāsaṅkhyāta*	Lowest higher order innumerable	$\alpha^{2\alpha}$
11	Madhyama *asaṅkhyātāsaṅkhyāta*	Intermediate higher order innumerable	$\alpha^{2\alpha} + 1, \ldots, \beta$-2
12	utkṛṣṭa *asaṅkhyātāsaṅkhyāta*	Highest higher order innumerable	$\beta - 1$
13	Jaghanya paritānanta	Lowest lower order infinity	β (as defined)
14	Madhyama paritānanta	Intermediate lower order infinity	$\beta + 1, \ldots, \beta^{\beta}$-2
15	utkṛṣṭa paritananta	Highest lower order infinity	β^{β}-1
16	Jaghanya yuktānanta	Lowest middle order infinity	β^{β}
17	Madhyama yuktānanta	Intermediate middle order infinity	$\beta^{\beta} + 1, \ldots, \beta^{2\beta}$-2
18	utkṛṣṭa yuktānanta	Highest middle order infinity	$\beta^{2\beta}$-1
19	Jaghanya anantānanta	Lowest higher order infinity	$\beta^{2\beta}$
20	Madhyama anantānanta	Intermediate higher order infinity	$\beta^{2\beta} + 1, \ldots, \Omega$-1
21	utkṛṣṭa anantānanta	Highest higher order infinity	Ω (as defined)

According to Jaina cosmology the universe is composed of alternate rings of land and sea whose boundaries are concentric circles with increasing radii. The width of any ring whether land or sea is double that of the preceding ring. The inner most circle is called Jambūdvīpa and has a diameter of 100000 yojans [24]

Consider four cylindrical pits of 100000 yojanas in diameter and 1000 yojanas deep, namely A_1, B_1, C_1, and D_1. A_1 is filled with mustard seeds of radius ½*sarṣapa* $= 1/(2 \times 6 \times 8 \times 8 \times 8 \times 8 \times 10^6)$ *yojans* and further seeds are pilled over it in the form of conical shape, the topmost layer consisting of one seed. The total number of seeds required for the operation is given by $N_0 = V/v$, where V is volume of the pit and v is volume of the seed. The value of N_0 is calculated as 1.979×10^{44}. Later on, the value of N_0 was determined by slightly different way as 2.2265×10^{44} by R. C. Gupta [25] and it was also observed by R. S. Shah [26] that Nemicandra used some erroneous formula to calculate the volume of spherical seed. The value of N_0 is very huge number. For the number of seeds in the next pit, whose diameter is equal to that of N_0^{th} sea is $(2^{(N0+1)} - 3) \times$ 100000 yojans. Volume of the pits are proportional to square of the diameter hence we can write $N_1 = (2^{(N0+1)} - 3)^2 \times N_0 \approx 2^{2(N0+1)} \times N_0 \approx 10^{10^{44}} = 10^{N0}$. From this we can see that N_0 is infinitesimal in comparison to N_1. With the same approximation $N_2 = 2^{2(N0+N1+1)} \times N_0 \approx 2^{2N1} \times N_1$ where $2^{2N1} \approx 10^{10^{10^{44}}}$.

Still, this number is very very less than Asaṅkhyāta. As mentioned by R. S. Shah the value of minimum lower order innumerable is given by $\alpha = N_0 + N_1 + N_2 + \cdots + N_{N0-1} + N_{N0}$. In the sum on the right side each term is infinitesimal than its successor term and can be ignored. Thus, $\alpha = N_{N0}$ [27].

5.3. Classification of Infinity

The *Dhavalā* [28] gives a classification of infinity. According to it, there are 11 kinds of infinity [29]:

1. **Namānanta**: Infinity in name. An aggregate of objects which may or may not really be infinite might be called as such,

in ordinary conversation, or by or for ignorant persons, or in literature to denote greatness. In such contexts the term infinite means infinity in name only.

2. **Sthāpanānanta:** Attributed, or associated infinity. This too is not the real infinity. The term is used in case infinity is attributed to or associated with some object.
3. **Dravyānanta:** Infinite in relation to knowledge which is not used. This term is used for persons who have knowledge of the infinite, but do not for the time being use that knowledge.
4. **Gaṇanānanta:** The numerical infinite. This term is used for the actual infinite as used in mathematics.
5. **Apradeśikānanta:** Dimensionless or infinitely small.
6. **Ekānanta:** One directional infinity. It is the infinite as observed by looking in one direction along a straight line.
7. **Ubhayānanta**: Two directional infinity. This is illustrated by a line continued to infinity in both directions.
8. **Vistārānanta:** Two dimensional or superficial infinity. This means as an infinite plane area.
9. **Sarvānanta:** Spatial infinity. This signifies the three-dimensional infinity i.e., the infinite space.
10. **Bhāvānanta:** Infinite in relation to knowledge which is utilised. This term is used for a person who has knowledge of the infinite, and who uses that knowledge.
11. **Śaśvatānanta:** Everlasting or indestructible.

Dhavalā clearly lays down that, in the subject matter under discussion, by the term ananta (infinite) we always mean the numerical infinite and not any of the other infinities enumerated above [30]. For in the other kinds of infinity the idea of enumeration is not found. It has also been stated that the numerical infinite it is describable at great length and is simpler. This statement probably means that in Jaina literature infinite was defined more thoroughly by different writers and had become commonly used and understood. The *Dhavalā*, however, does not contain a definition of ananta. On the other hand, operations on and with the *ananta* are frequently mentioned along with numbers called *saṅkhyāta* and *asaṅkhyāta*. The number *saṅkhyāta*, *asaṅkhyāta* and *ananta* have been used in Jaina literature from the earliest known times, but it seems they did not always carry the same meaning. In the earlier literature *ananta* was certainly used in the sense of infinity as we understand it now, but in later works *anantānanta*, takes the place of *ananta*. For instance, in *Trilokasāra*, *parita-ananta*, *yuktānanta* and even *jaghanya anantānanta* is very big number but is finite. According to this work, *utkṛṣṭa anantānanta* is ultimate infinity [31].

5.4. Asaṅkhyāta and Ananta

Asaṅkhyāta is a finite number which is exhausted by continuous subtraction of one after one [32]. On the other hand if it never ends then it is an *ananta*. *Asaṅkhyāta* literally means innumerable and can be visualized as finite number that is not considered in number naming scheme [33]. It could be a number which is outside the scope of elementary cognitive sphere in the sense that it can not be expressed numerically [34]. Nemicandra defined *asaṅkhyāta* as a number equal to 10^{140} [35]. *Ananta* is a transfinite number but such a clarity does not exist about *asaṅkhyāta*. Navjyoti Singh [36] in his article pointed about confusion regarding finitude and infinitude of *asaṅkhyāta*. As per his findings, *asaṅkhyāta* signified infinity only when *asaṅkhyāta* space points and *asaṅkhyāta* instants of time would mean actual physical infinity and hence provide a consistent cosmological picture. The Jaina literature may have had a physical notion of actual infinity from the cosmological point of view, but mathematically they did not have such a notion of *asaṅkhyāta*.

Moreover, even in *Dhavalā* where finitude of *asaṅkhyāta* is categorically stated, similarity of various meanings of the two classes of numbers is also discussed. Identical eleven-fold classification of various meanings of *ananta* and *asaṅkhyāta* is given in *Dhavalā*. Among eleven kinds of meaning the comparison of space less (*apradeśi*) *ananta* and space less (*apradeśi*) *asaṅkhyāta* gives interesting result. *Apradeśikānanta* refers to one monad of *pudgala* and *apradeśikāsaṅkhyāta* refers to one *pradeśa* of soul. Here both *ananta* and *asaṅkhyāta* denote the concept of actual infinity. Out of the eleven-fold classification only *gaṇana* is open to numeration and to possible meaning of mathematical actual infinity. Apart from this classification identical nine-fold division of *ananta* and *asaṅkhyāta* is also given in *Dhavalā*. Therefore, on the basis of parallelism between these two, we can not consider *asaṅkhyāta* as transfinite number or mathematical actual infinity. Navjyoti singh [37] in his work has rejected A. N. Singh's argument that *ananta pradeśas* of *pudagla* or *ananta pradeśas* of *lokākāśa* are transfinite number but *asaṅkhyāta pradeśas* of *lokākāśa* or *saṅkhyāta pradeśa* of a soul are not transfinite. In his article, he demonstrated that *asaṅkhyāta* and *ananta* were used in the sense of transfinite numbers in Jaina literature.

5.5. Conclusion

Jaina developed the concept of transfinite theory before 800 AD while the basic ordering of physical actual infinities could have been developed as early as 300 BC. Jaina ascetic mathematicians were the earliest to give the broad classification of numbers by dividing them into three categories- *saṅkhyāta*, *asaṅkhyāta* and *ananta*. There is no difference between *asaṅkhyāta* and ananta class of numbers. The confusion was due to two different schemes.

Acknowledgement

I would like to express my sincere gratitude towards Dr. Anupam Jain (Director, Center for Studies of Ancient Indian Mathematics) Devi Ahilya Vishwavidyalaya, Indore who not only suggested to work on this topic but also provided the related literature available with him to carry out the work. I am also thankful to Dr. Pragati Jain, who holds the doctoral degree on mathematics of *Dhavala*. She helped me to share her view on the infinity as described by Virasena. I am thankful to referees for their valuable comments and suggestions. Last but not least I want to convey my thanks to Dr. S. K. Jain for continuous follow ups and guidance throughout the paper.

References

[1] G.G. Joseph, The Crest of The Peacock Princeton University Press, Princeton, 2000, p. 351.

[2] Brahmagupta, *Brahmasphuṭasidhhānta*, Edited with Sanskrit and Hindi commentaries by Ram Swarup Sharma and His team, Indian Institute Astronomical and Sanskrit Research, New Delhi, 1966. Vol. III P. Verses 18. 30–35.

[3] Kim Plofker, Mathematics in India, Princeton University Press, Princeton, 2009, p. 151.

[4] ibid, p. 192.

[5] Mahāvīra, *Gaṇitasārasaṅgraha*, Edited with an English translation by M. Rangacharya, Madras, 1912. Edited and translated by L. C. Jain into Hindi, Jivaraja Granthamala No. 12, Jaina Sanskrti Samraksaka Sangha, Sholapur, 1963. p. 491.

[6] A. N. Singh, Mathematics of *Dhavalā*, Article published with *Dhavalā*, Amaraoti. Reprinted Jaina Samskrti Samraksaka Sangha, Sholapur, 1996. p. 314.

[7] Jaini, Padmanabh S, Gender and Salvation: Jaina Debates on the Spiritual Liberation of Women, University of California Press, ISBN 0-520-06820-3, 1991.

[8] Indranandi, *Shrutavatara*.

[9] A. N. Singh, Mathematics of *Dhavalā*, p. 302.

[10] Vīrasena, *Dhavalā*, a commentary on *Śaṭkhaṇḍāgama* of Puṣpadanta and Bhūtabalī, edited and translated by H. L. Jain, A. N. Upadhye and K.S. Shastri, Rev. Edition, Jain Sanskrit Samrakshak Sangh, 1980. Volume III, p. 98–100.

[11] ibid, p. 53.

[12] ibid, p. 200.

[13] ibid, p. 253.

[14] ibid, p. 21.

[15] R.S. Shah, Jaina Mathematics: Lore of Large Numbers, Bulletin of Marathwada Mathematical Society, Aurangabad, Vol-10, No.-1, June 2009.

[16] *Palyopama*, is an inestimably long period of time. It is calculated as follows: a vessel, a *yojana* wide and deep, is filled with the hairs of a new-born lamb—hairs that have grown within seven days. If one hair is withdrawn every hundred years, the time required to empty the vessel is a *palyopama*. —Cf. commentary to *Tattvārthadhigamasūtra* 4. 15.

[17] *Sāgaropama* refers to a unit of time equaling ten crores of *palyopamas*, which is calculated as follows: a vessel, a *yojana* wide and deep, is filled with the hairs of a new-born lamb—hairs that have grown within seven days. If one hair is withdrawn every hundred years, the time required to empty the vessel is a *sāgaropama*. —Cf. commentary to *Tattvārthadhigamasūtra* 4. 15.

[18] Vīrasena, *Dhavalā* (III, p. 18,126).

[19] Aryaraksita, *Anuyogadvara Sutra, Edited by* Misrimalajl Maharaja 'Madhukara'and his editorial team, Publication No. (Granthanka) 28, Sri Agama Prakasana Samiti, Beawar (Rajsthan), 1987.sutra 497-518

[20] Vīrasena, *Dhavalā*, (III, p. 18,126).

[21] A. N. Singh, Mathematics of *Dhavalā*, p. 310. R. S. Shah, Jaina Mathematics: Lore of Large Numbers, p. 51.

[22] R.C. Gupta, Ancient Jaina Mathematics, p. 225.

[23] Nemicandra Siddhāntacakravartī, *Trilokasāra*, Bombay, 1920, edited with Hindi commentary, Aryika Viśuddhamati, Sahitya Bharti Prakashan, Gannaur, Sonipat, 2006, P. 30–35, *gatha* 17–23.

[24] Padamanandi, *Jambūdīva-paṇṇatti-saṃgaho*, edited with Hindi translation and notes by H. L. Jain & A. N. Upadhye, Jain Sanskriti Samrakshaka Samgha, Sholapur, 1958, reprint 2004, *Uddesa* 2, *gatha* 1–15.

[25] R. C. Gupta, The First Unenumerable Number in Jaina Mathematics, Gaṇita Bhāratī, Volume 14 (1-4) 1992, p. 11–24.

[26] R.S. Shah, Jaina Mathematics: Lore of Large Numbers, p. 52.

[27] Ibid. p. 54.

[28] Vīrasena, *Dhavalā*, a commentary on *Śaṭkhaṇḍāgama* of Puṣpadanta and Bhūtabalī, edited and translated by H. L. Jain, A. N. Upadhye and K. S. Shastri, Rev. Edition, Jain Sanskrit Samrakshak Sangh, 1980.

[29] Ibid. (III, p. 11–16).

[30] *Ibid p. 16.*

[31] A. N. Singh, Mathematics of *Dhavalā*, p. 314.

[32] Vīrasena, *Dhavalā* (III, p. 25).

[33] A. N. Singh, Mathematics of *Dhavalā*, p. 309.

[34] Vīrasena, *Dhavalā* (III, p. 268).
[35] A. N. Singh, Mathematics of *Dhavalā*, p. 312.
[36] Navjyoti Singh, Jain, Theory of Measurement and Theory of Transfinite Numbers, Proceeding of International Seminar on Jaina Mathematics & cosmology, DJICR, Meerut,1991, p. 210.
[37] Ibid p. 217.

https://doi.org/10.1142/9789811255502_0006

Article 6

Geometry in Mahāvīrācārya's *Gaṇitasārasaṅgraha*

S. Balachandra Rao

#2388, JnanaDeepa 13th Main, A-Block
Rajajinagar 2nd Stage, Bangalore 560 010
balachandra1944@gmail.com

K. Rupa

Department of Mathematics, Global Academy of Technology
Rajarajeshwari Nagar, Bangalore 98, India
rupak@gat.ac.in

S.K. Uma*

Department of Mathematics, Sir Mokshagundam Visvesvaraya
Institute of Technology, Bangalore 560 157, India
uma_mca@sirmvit.edu

PadmajaVenugopal

Department of Mathematics, SJB Institute of Technology
Bangalore 60, India
venugopalpadmaja@gmail.com

The Mathematical genius of the ancient Indians was mainly computational, but the inspiration is geometry. The main developments in Indian mathematics were oriented and inspired by the needs of astronomy. The branch of Mathematics which received earliest attention was geometry. Geometry is designated as *kṣetragaṇita* in most Indian mathematical works, *Kṣetra* means a closed figure. Also called *rajju* or *rajjugaṇita*

*Corresponding author.

(calculation with the cord) in the *Śulbasūtras*. In Jaina literature there are four *anuyogās* called *prathama*, *karaṇa*, *caraṇa* and *dravya*. In *karaṇānuyoga*, many mathematical operations are used. The crowning glory of the succession of the great Jaina mathematicians since ancient times, was the renowned Jaina mathematician, Mahāvīrācārya (c. 9th century CE) from Karnataka, who flourished in the royal court of the famous Rāṣṭrakūṭa king Amoghavarṣa Nṛupatuṅga (815–878 CE). Mahāvīrācārya not only made a very systematic compilation of the results of his predecessors but also made many original and significant contributions to mathematics. His work *Gaṇitasārasaṅgraha* ("Compendium of the Essence of Mathematics") is the first independent mathematical text and not part of any astronomical work. In this paper we present Mahāvīrācārya's work on the areas, circumferences, volumes, etc. of the usual geometrical figures as well as certain special curves. They include triangular shapes, quadrilateral shapes, circular shape, ring shapes and some interesting combinations of these.

Keywords: Mahāvīrācārya, *Gaṇitasārasaṅgraha*, *Kṣetragaṇita*, *Rajjugaṇita*, *Rekhāgaṇita*.

Mathematics Subject Classification 2020: 01-06, 01A32, 01A99

6.1. Introduction

The subject of Mathematics was given high importance in India since the Vedic period itself. Mathematics(*Gaṇita*) along with Astronomy was included in *Jyotiṣa*. Geometry was developed in the *Śulbasūtras*. in the first millennium BCE in the context of construction of vedis for citis, and later in the Siddhānta astronomy tradition pioneered by Āryabhaṭa (b. 476). In this paper we present an exposition of geometry in the renowned work *Gaṇitasārasaṅgraha* (GSS, in the sequel) of Mahāvīrācārya (ca. 850 CE); see [1] for reference.

The 7th chapter of *GSS* is completely dedicated to geometry. The chapter comprises of 232 versified rules and examples. In this chapter Mahāvīrācārya discusses geometrical shapes like trilateral, quadrilateral, curvilinear figures and the methods to find their areas. He also gives rules or formulae for computing the areas of figures which are not basic geometric figures, but which can be composed from basic geometric figures called *yavā* (barley corn), *muraja* (a sort of drum) and *śaṅkha* (conch shell). Mahāvīrācārya was the first Indian Mathematician to discuss these shapes:

Yavākārakṣētra, *Murajākārakṣētra*, *Paṇavākārkṣētra* and *Vajrākārakṣētra*.

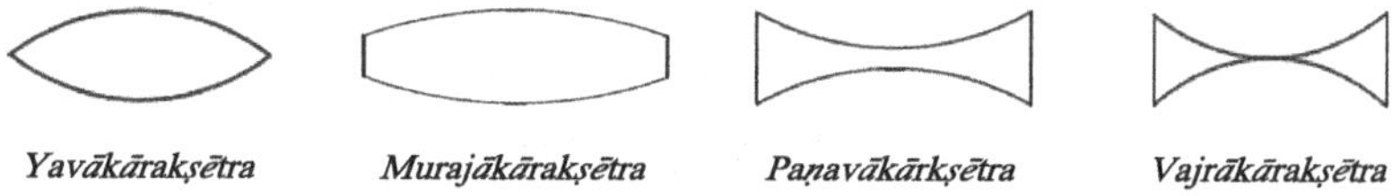

Figure 6.1.

Mahāvīrācārya enumerates eight varieties of curvilinear figures. His classification of quadrilaterals differs from that of Śrīdhara, who takes the primary plane figures to be ten in number.

6.2. Trilaterals and Quadrilaterals

Three types of trilateral are discussed by Mahāvīrācārya. These are: Equilateral triangles, Isosceles triangles and Scalene triangles.

Mahāvīrācārya considers a triangle as a quadrilateral of one side zero and gives the following rule to find its area:

भुजयुत्यर्धचतुष्काद्भुजहीनाद्घातितात्पदं सूक्ष्मम् ।
अथवा मुखतलयुतिदलमवलम्बगुणं न विषमचतुरश्रे ॥

bhujayutyardhacatuṣkādbhujahīnāddhātitātpadaṁ sūkṣmaṁ
athavā mukhatalayutidalamavalambaguṇaṁ na viṣamacaturaśre.

(7 *ś*/50)

In this *śloka* Mahāvīrācārya says that from the half of the sum of four sides of a quadrilateral subtract each side (separately). Finding the product of such quantities and then taking the square root, the minutely accurate measure of the area of the quadrilateral is obtained or the product of the perpendicular from the top to the base and half the sum of the top measure and the base measure will give the minutely accurate area of a quadrilateral

$$\sqrt{(s-a)(s-b)(s-c)(s-d)}$$

where a, b, c, d are the four sides and $s = \frac{a+b+c+d}{2}$.

In the case of a trilateral, choosing d to be 0 we have $\sqrt{s(s-a)(s-b)(s-c)}$ and $s = \frac{a+b+c}{2}$.

Also Area $= \frac{c}{2} \times p$ where c is one side of the triangle and 'p' is the perpendicular distance of the opposite vertex from this side.

भुजकृत्यन्तरभूहृतभूसङ्क्रमणं त्रिबाहुकाबाधे।
तद्भुजवर्गान्तरपदमवलम्बकमाहुराचार्याः ।।

bhujakṛtyantarabhūhṛtabhūsaṅkramaṇaṁ tribāhukābādhe,
tadbhujavargāntarapadamavalambakamāhurācāryāḥ.

(7 *ś*/49)

In this *śloka* it is given that by doing *sankramaṇa* between one side base and the difference between the squares of the other sides as divided by the base will give the two segments of the base. The square root of the difference between the 3 squares of these segments and of the corresponding adjacent sides gives the measure of the perpendicular. Let

$$c_1 = \frac{1}{2}\left(c + \frac{a^2 - b^2}{c}\right) \quad \text{and} \quad c_2 = \frac{1}{2}\left(c - \frac{a^2 - b^2}{c}\right).$$

Then $p = \sqrt{a^2 - c_1^2}$ or $\sqrt{b^2 - c_2^2}$.

The product of the perpendicular (from the vertex to the base) and half the base gives the measure of the area of a triangle i.e. $\frac{1}{2} \times$ base $\times$ altitude. Āryabhaṭa gives this formula for area of a trilateral; see [3, 4]. Bhāskara I also elucidates Āryabhaṭa's expression for area of a triangle.

Brahmagupta also gave the formula for the same as $\sqrt{s(s-a)(s-b)(s-c)}$ where a, b, c are the sides and $s = \frac{a+b+c}{2}$ treating triangle as a quadrilateral with one side 0.

In-circle of a triangle

In the Indian context Mahāvīrācārya is the first to speak of a circle inscribed within a triangle and its diameter.

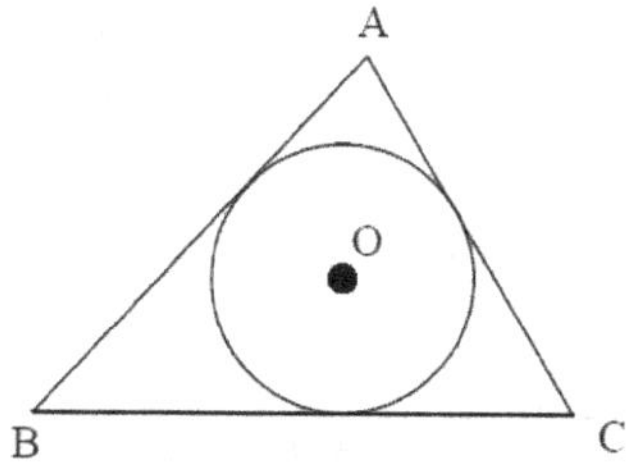

Figure 6.2.

परिधेः पादेन भजेदनायतक्षेत्रसूक्ष्मगणितं तत् ।
क्षेत्राभ्यन्तरवृत्ते विष्कम्भोऽयं विनिर्दिष्टः ।।

paridheḥ pādenā bhajedanāyatakṣetrasūkṣmagaṇtiaṁ tat,
kṣetrabhyantarāvṛtte viṣkambho'yaṁ vinirdiṣṭaḥ.

(7 *śl*223 1/2)

The above verse says that the exact area of any figure other than a rectangle should be divided by one-fourth the perimeter to get the diameter of the circle inscribed within that figure.

In a triangle ABC, the circle inside the triangle, touching all the sides is called the in-circle of triangle ABC. GSS gives a formula for the diameter of the in-circle as

$$\text{In-diameter} = \frac{\text{Area of triangle ABC}}{\frac{1}{4} \times \text{circumference of triangle ABC}}.$$

Mahāvīrācārya discusses five varieties of quadrilaterals in *śloka*'s 54–59 namely

(i) *Samacaturaśra*(equilateral quadrilateral)
(ii) *Āyatacaturaśra* (longish quadrilateral)
(iii) *Trisamacaturaśra*(equitrilateral quadrilateral)
(iv) *Dvisamacaturaśra* (equilateral quadrilateral)
(v) *Viṣamacaturaśra*(inequilateral quadrilateral)

Mahāvīrācārya in *śloka* 50 gives the formula for area of a quadrilateral, but notes that this expression is not applicable to *viṣamacaturaśra*; he thus seems to have been aware of some limitations in the applicability of the formula, though there are no further details given.

Mahāvīrācārya gives a method of obtaining a quadrilateral with a given (integer) area.

धनकृतिरिष्टच्छेदैश्चतुर्भिराप्तैव लब्धानाम् ।
युतिदलचतुष्टयं तैरूना विषमाख्यचतुरश्रभुजसङ्ख्याः ।।

dhanakṛtiriṣṭacchedaiścaturbhirāptaiva labdhānām,
yutidalacatuṣṭayaṃ tairūnā viṣamākhyacaturaśrabhujasaṅkhyā.

(7 ś/152)

The rule given here stipulates the numerical value of the area to be squared and then written as a product of four factors. Suppose the area of a quadrilateral is given as an integer Δ. Find four integer factors α, β, γ and δ of Δ^2 such that $\Delta^2 = \alpha\beta\gamma\delta$; though this has not been specified in GSS, it can be seen that the factors can be chosen so that the sum $\alpha + \beta + \gamma + \delta$ is even. Now let $s = \frac{(\alpha+\beta+\gamma+\delta)}{2}$ then s is an integer and the four required sides a, b, c and d are $(s - \alpha)$, $(s - \beta)$, $(s - \gamma)$, $(s - \delta)$ respectively. The reader is referred to [2], Chapter III, Section 20, for Mahāvīrācārya's treatment of rational rectangles.

6.3. Circles

The area and circumference are described by Mahāvīrācārya, giving approximate and accurate value as follows:

The approximate area of a circle is given by

त्रिगुणीकृतविष्कम्भः परिधिर्व्यासार्धवर्गराशिरयम् ।
त्रिगुणः फलं समेऽर्धे वृत्तेऽर्धं प्राहुराचार्याः ।।

triguṇīkṛtaviṣkambhaḥ paridhirvyāsārdhavargarāśirayam,
triguṇaḥphalaṁsame'rdhe vṙtte'rdhaṁ prāhurācāryāḥ.

(7 ś/19)

The above *śloka* mentions that for a circle the measure of the diameter multiplied by 3 is the measure of the circumference and the number representing the square of half the diameter multiplied by 3 is the area, which means approximate area $= 3 \times \left(\frac{d}{2}\right)^2$ and approximate perimeter $= 3 \times$ diameter (here he has approximated the value of π is 3).

In *śloka* 60 Mahāvīrācārya gives the rule for obtaining minutely accurate values relating to curvilinear figures:

वृत्तक्षेत्रव्यासो दशपदगुणितो भवेत्परिक्षेपः।
व्यासचतुर्भागगुणः परिधिः फलमर्धमर्धे तत् ।।

vṛttakśetravyāso daśapadaguṇito bhavetparikṣ epaḥ
vyāsacaturbhāgaguṇaṇ paridhiḥ phalamardhamardhe tat.

(7 *ś*/60)

The diameter of the circular figure multiplied by the square root of 10 becomes the circumference (in measure). The circumference multiplied by one-fourth of the diameter gives the area. In the case of a semi-circle this happens to be half (of what it is in the case of the circle).

Hence circumference $c = \sqrt{10}d$, and area $= c \times \frac{d}{4} = \sqrt{10}\left(\frac{d}{2}\right)^2$.

The value of π is approximated to $\sqrt{10}$ in this stanza, which is equal to 3.16.......

Area of a segment

इषुपादगुणश्च गुणो दशापदगुणितश्च भवति गणितफलम् ।
यवसंस्थानक्षेत्रे धनुराकारे च विज्ञेयम् ।।

iṣupādaguṇaśca guṇo daśapadaguṇitaścabhavati gaṇitaphalaṁ
yavasaṁsthānakṣetre dhanurākāre ca vijñeyaṁ

(7 *śl* 70 1/2)

Figure 6.3.

The above *śloka* says the measure of the string(chord) multiplied by one-fourth of the measure of the arrow and then multiplied by the square root of 10 gives the accurate area of a figure having the outline of a bow and also in the case of a figure resembling the longitudinal section of a *yava* grain. Thus

Area of segment $= \sqrt{10}\frac{jiva}{4} \times bana = \sqrt{10}\frac{cp}{4}$.

The following *śloka*

कृत्वेषुगुणसमासं बाणार्धगुणं शरासने गणितम् ।
शरवर्गात्पञ्चगुणाज्ज्यावर्गयुतात्पदं काष्ठम् ।।

kṛtveṣuguṇasamāsaṁ baṇārdhaguṇaṁ śarāsane gaṇitaṁ

śaravargātpañcaguṇājjyāvargayutātpadaṁ kāṣṭhaṁ.

(7 *śl* 43)

translates as, the area of a field resembling a bow in outline is in fact the segment of a circle, the bow forming the arc, the bow-string forming the chord and the arrow measuring the greatest perpendicular distance between the arc and the chord.

This means the square root of the square of the arrow as multiplied by five and then combined with the square of the measure of the string gives the approximate measure of the bow. Thus

Length of bow $= \sqrt{5p^2 + c^2}$.

Length of arrow $= \frac{\sqrt{a^2 - c^2}}{5}$.

Length of bow-string $= \sqrt{a^2 - 5p^2}$.

Area $= (c + p) \times \frac{p}{2}$.

More accurate values can be seen in the following verse

शरवर्गः षड्गणितो ज्यावर्गसमन्वितस्तु यस्तस्य ।
मूलं धनुर्गुणेषुप्रसाधने तत्र विपरीतम्: ।।

śaravargaḥ ṣaḍguṇito jyāvargasamanvitastu yastasya,

mūlaṁ dhanurguṇeṣuprasādhane tatra viparītaṁ.

(7 *śl* 73 1/2)

The square of the arrow measure is multiplied by 6. To this is added the square of the string measure. The square root of that (the resulting sum) gives the measure of the bow stick. For finding the measure of the string and the measure of the arrow, a course converse to this is adopted.

Length of bow $= \sqrt{6p^2 + c^2}$,

Length of arrow $= \frac{\sqrt{a^2-c^2}}{6}$,

Length of bow-string $= \sqrt{a^2 - 6p^2}$.

6.4. Areas of Some Special Curves

One of the very significant contribution of Mahāvīrācārya is a discussion of some special curves, like the shapes of ***Mṛdaṇga***, ***Paṇava*** and ***Vajra***. The formula described in case of these are only approximate. It was probably from the point of view of some practical utility that all the results have been stated separately.

Rule for arriving at the practically approximate value of surface areas of figures resembling Mṛdaṅga, *Paṇava* and *Vajra*

यवमुरजपणवशक्रायुधसंस्थानप्रतिष्ठितानां तु ।
मुखमध्यसमासार्धं त्वायामगुणं फलं भवति ।।

yavamurajapaṇavaśakrāyudhasaṃsthānapratiṣṭitānāṁ tu,
mukhamadhyasamāsārdhaṁ tvāyāmaguṇaṁ phalaṁ bhavati.

(7 *śl*32)

In the case of areas shaped in the form of the *yava* grain, of the *muraja*, of the *paṇava* and of the *vajra*, the (required measure of) area is that which results by multiplying half the sum of the end measure and the middle measure by the length.

मुखगुणितायामफलं स्वधनुःफलसंयुतं मृदङ्गनिभे ।
तत्पणववज्रनिभयोर्धनुःफलोनं तयोरुभयोः ।।

mukhaguṇitāyāmaphalaṁ svadhanuḥphalasaṁyutaṁ mṛdaṅganibhe,
tatpaṇavavajranibhayordhanuḥphalonaṁ tayorubhayoḥ.

(7 *śl* 76 1/2)

To the resulting area, obtained by multiplying the (maximum) length with (the measure of the breadth of) the face, the value of the areas of its associated bow- shaped figures is added. The resulting sum gives the value of the area of a figure resembling (the longitudinal section of) a *mṛdaṅga*. In the case of those two (other) figures which resemble (the longitudinal section of) the *paṇava* and (of) the *vajra*, that (same resulting area, which is obtained by multiplying the maximum length with the measure of the breadth of the face), is diminished by the measure of the areas of the associated bow-shaped figures.

Mṛdaṅga

चतुर्विंशतिरायामो विस्तारोऽष्टौ मुखद्वये ।
क्षेत्रे मृदङ्गसंस्थाने मध्ये षोडश किं फलम् ।।

caturviṁśatirāyāmo vistāro'ṣṭau mukhadvaye,
kṣetre mṛdaṅgasaṁsthāne madhye ṣoḍaśa kiṁ phalaṁ.

(7 *śl* 771/2)

In the case of a figure having the outline configuration of a *Mṛda nga*, the (maximum) length is 24; the breadth of (each of) the two facesis 8 and the (maximum) breadth in the middle is 16. What is the area?

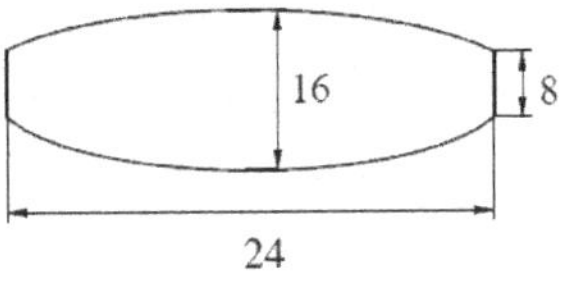

Figure 6.4.

length $\times$ breadth $= 24 \times 8 = 192$.
Area of bow shaped segment $= \sqrt{10}\left(\frac{24\times4}{4}\right) = \sqrt{5760}$.
Therefore required Area $= 192 + \sqrt{5760} + \sqrt{5760} = 192 + \sqrt{23040}$.

Paṇava

चतुर्विंशतिरायामस्तथाष्टौ मुखयोर्द्वयोः ।
चत्वारो मध्यविष्कम्भः किं फलं पणवाकृतौ ।।

caturviṁśatirāyāmastathāṣṭau mukhayordvayoḥ, catvāro madhyaviṣkambhaḥ kiṁ phalaṁ paṇavākṛtau.

(7 *śl* 78 1/2)

In the case of a figure having the outline of a *Paṇava*, the (maximum) length is 24; similarly, the measure (of the breadth of either) of the two faces is 8 and the central breadth is 4. What is the area?

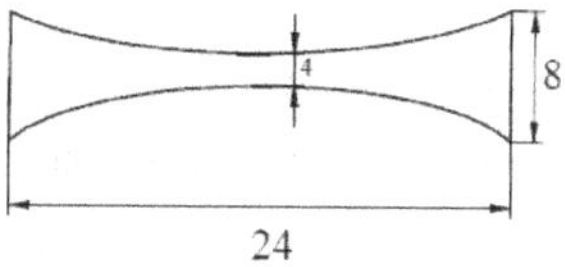

Figure 6.5.

length $\times$ breadth $= 24 \times 8 = 192$.
Area of bow shaped segment $= \sqrt{10}\frac{24\times 4}{4} = \sqrt{5760}$.
Therefore required Area $= 192 - \sqrt{5760}$.

Vajra

चतुर्विंशतिरायामस्तथाष्टौ मुखयोर्द्वयोः ।
मध्ये सूचिस्तथाचक्ष्व वज्राकारस्य किं फलम् ।।

caturviṁśatirāyāmastathāṣṭau mukhayordvayoḥ, madhye
sūcistathācakṣva vajrākārasya kiṁ phalaṁ.

(7 *śl* 79 1/2)

In the case of a figure having the outline of a *Vajra*, the (maximum) length is 24; the measure (of the breadth of either) of the face is 8 and the centre is a point. Give out as before what the area is given by

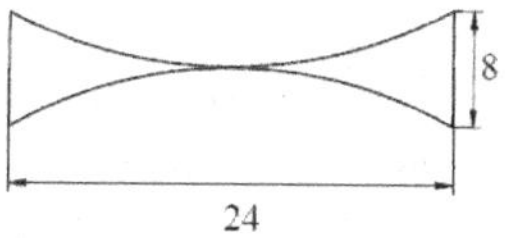

Figure 6.6.

length $\times$ breadth $= 24 \times 8 = 192$.
Area of bow shaped segment $= \sqrt{10}\frac{24\times 4}{4} = \sqrt{5760}$.
Therefore required Area $= 192 - \sqrt{5760} - \sqrt{5760} = 192 + \sqrt{23040}$.

6.5. Area Bounded by Circles

Mahāvīrācārya is apparently the first mathematician to have discussed the formulas for the area of the space enclosed by 3 or more equal and mutually touching circles.

विष्कम्भवर्गराशेर्वृत्तस्यैकस्य सूक्ष्मफलम् ।
त्यक्त्वा समवृत्तानामन्तरजफलं चतुर्णां स्यात् ।।

viṣkambhavargarāśervṛttasyaikasya sūkṣmaphalaṁ tyaktvā
samavṛttānāmantarajaphalaṃ caturṇāṁ syāt.

(7 *śl* 82 1/2)

In this *śloka* Mahāvīrācārya describes the method to find the minutely accurate area of the region enclosed by **four circles** touching each other. By subtracting the minutely accurate area of a circle from the square of the diameter the area of the inter space including within four equal circles touching each other is obtained. I.e., if d is the diameter of each circle then

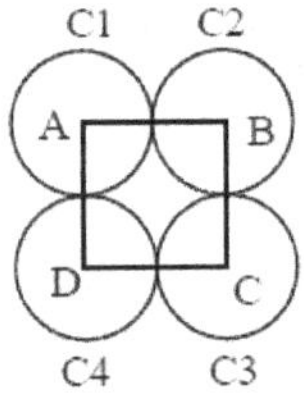

Figure 6.7.

$A =$ (Diameter square) $-$ (Area of one circle) or $A = d^2 - \frac{\sqrt{10}d^2}{4}$.

Where the equilateral triangle is formed by the centers of the **three circles**. Each side of the triangle is equal to the diameter of each of the circle.

If d is the diameter of each circle, the triangle formed by the centers of these circles as vertices is equilateral with each side d. then the area of the region bounded by these circles

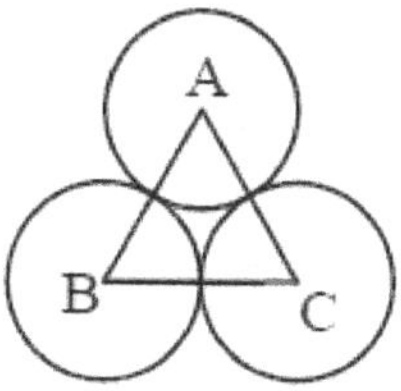

Figure 6.8.

$$A = \frac{\sqrt{3}}{4}d^2 - \frac{1}{2}\left(\sqrt{10}\frac{d^2}{4}\right).$$

Mahāvīrācārya generalises his analysis to get approximate area for the space enclosed by any number of equal circles touching each other as

रज्ज्वर्धकृतित्र्यंशो बाहुविभक्तो निरेकबाहुगुणः ।
सर्वेषामश्रवतां फलं हि बिम्बान्तरे चतुर्थांशः ।।

rajjvardhakṛtitryaṁśo bāhuvibhakto nirekabāhuguṇaḥ,
sarveṣāmaśravatāṁ phalaṁ hi bimbāntare caturthāṁśaḥ.

(7 *śl*39)

which means that one third the square of the semi perimeter divided by the number of sides and multiplied by the number of sides diminished by one gives the area of all regular figures and one fourth of this is the area enclosed between the circles.

i.e. the area of a polygon of n sides each equal to A is $\frac{s^2}{3}\left(\frac{n-1}{n}\right)$ where s is the semi perimeter. Then the area enclosed by the circles at the vertices of the regular polygon $= \frac{1}{4}\left(\frac{s^2}{3}\right)\left(\frac{n-1}{n}\right)$.

6.6. Elongated Circle (*Āyatavṛtta*)

Although Mahāvīrācārya's formula for the area of an ellipse is wrong, his formula for circumference $\sqrt{16a^2 + 24b^2}$ reduces to $2\pi a\sqrt{1 - \frac{3}{5}e^2}$ which is a remarkably close approximation to the circumference of an ellipse.

$$\sqrt{16a^2 + 24b^2} = \sqrt{16a^2 + 24(a^2 - a^2e^2)}$$

$$i.e.\ \sqrt{40a^2 - 24a^2e^2} = 2\sqrt{10a^2 - 6a^2e^2} = 2a\sqrt{10}\sqrt{1 - \frac{3}{5}e^2}$$

$$i.e.\ 2a\pi\sqrt{1 - \frac{3}{5}e^2}.$$

The circumference of an ellipse does not have a simple formula, unlike a circle ($2\pi r$).

It is significant to note that Mahāvīrācārya is the only Indian mathematician who discusses the approximate circumference of an ellipse. In fact, his approximation compares well with those given several centuries later.

6.7. Conchiform

Mahāvīrācārya gives the approximate area and the perimeter of the conchiform as

वदनार्धोनो व्यासस्त्रिगुणः परिधिस्तु कम्बुकावृत्ते ।
वलयार्धकृतित्र्यंशो मुखार्धवर्गत्रिपादयुतः ।।

vadanārdhono vyāsastriguṇaḣ paridhistu kambukāvṛtte,
valayārdhakṛtitryaṁśo mukhārdhavargatripādayutaḥ.

(7 *śl*23)

Figure 6.9.

The above verse means the maximum measure of the breadth diminished by half the measure of the face and multiplied by the square root of 3 gives the measure of the perimeter. One third of the square of half the perimeter increased by three-fourths of the square of half the measure of the face gives the area.

If d is the diameter and m is the face of a conchiform

$$Perimeter\ c = \sqrt{3}\left(d - \frac{1}{2}m\right) \quad \text{and} \quad Area = \left(\frac{C}{2}\right)^2 \times \frac{1}{3} + \frac{3}{4}\left(\frac{m}{2}\right)^2$$

Minutely accurate area and perimeter of conchiform

वदनार्धोनो व्यासो दशपदगुणितो भवेत्परिक्षेपः ।
मुखदलरहितव्यासार्धवर्गमुखचरणकृतियोगः ।।
दशपदगुणितः क्षेत्रे कम्बुनिभे सूक्ष्मफलमेतत्: ।।

vadanārdhono vyāso daśapadaguṇito bhavetparikṣepaḥ,
mukhadalarahitavyāsārdhavargamukhacaraṇakṛtiyogaḥ.
daśapadaguṇitaḥ kṣetre kambunibhe sūkṣmaphalametat.

(7 *śl 65 and 65* 1/2)

The (maximum measure of the) breadth (of the figure), diminished by half (the measure of the breadth) of the face and (then) multiplied by the square root of 10 gives rise to the measure of the perimeter. The square of half the (maximum) breadth as diminished by half the (breadth of the) face and the square of one-fourth (breadth of the) face are added together and the resulting sum is multiplied by the square root of 10.

This gives rise to the minutely accurate measure of the area in the case of the conchiform figure.

i.e. Area $= \left\{\left[\left(d - \frac{1}{2}m\right)\frac{1}{2}\right]^2 + \left(\frac{m}{4}\right)^2\right\}\sqrt{10}$ where d is the diameter and m is measure of the face.

Circumference $= \left(\mathrm{d} - \frac{\mathrm{m}}{2}\right)\sqrt{10}$.

6.8. Hexagon

भुजभुजकृतिकृतिवर्गा द्वित्रित्रिगुणा यथाक्रमेणैव।
श्रुत्यवलम्बककृतिधनकृतयश्च षडश्रके क्षेत्रे ।।

bhujabhujakṛtikṛtivargā dvitritriguṇā yathākrameṇaiva,
śrutyavalambakakṛtidhanakṛtayaśca ṣaṇaśrake kṣetre.

(7 *śl* 86 1/2)

It means in the case of a six-sided figure, the measure of the side, the square of the side, the square of the square of the side multiplied respectively by 2, 3 and 3 give rise, in that same order, to the values of the diagonal, of the square of the perpendicular and of the square of the measure of the area.

If a is the side of a hexagon, then

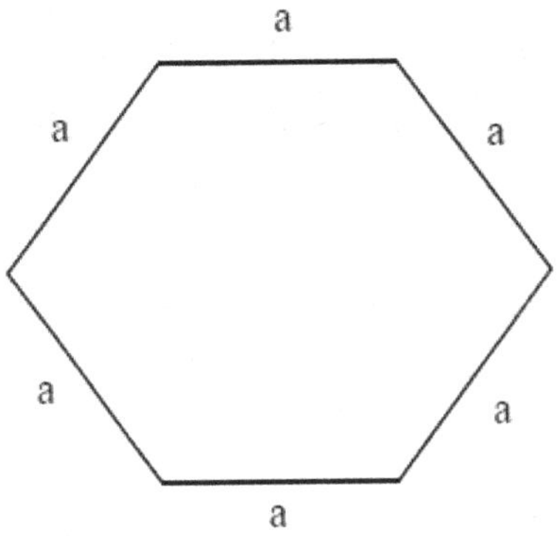

Figure 6.10.

$$\text{Diagonal} = 2a, \ \text{Perpendicular} = \sqrt{3a^2}, \ \text{Area} = \sqrt{3a^4}.$$

Here the hexagon meant to be a regular hexagon, though it is not mentioned to be so.

6.9. Sphere

Mahāvīrācārya gives the rule for finding approximate and accurate volume of a sphere as,

व्यासार्धघनार्धगुणा नव गोलव्यावहारिकं गणितम् ।

तद्दशमांशं नवगुणमशेषसूक्ष्मं फलं भवति ।।

vyāsārdhaghanārdhaguṇā nava golavyāvahārikaṁ gaṇitaṁ

taddaśamāṁśaṁ navaguṇamaśeṣasūkṣmaṁ phalaṁbhavatihḥ.

(8 *śl* 28 1/2)

i.e. approximate volume of a sphere is half of the cube of radius multiplied by 9. This approximate volume multiplied by nine and divided by 10,

Approximate volume $= \frac{9}{2}r^3$ where r is the radius of the sphere
Accurate volume $= \frac{9}{10} \times \frac{9}{2}r^3 = 4.05r^3$

Āryabhaṭa (see [3, 4]) gives the same as $= \frac{\pi d^2}{4}\sqrt{\frac{\pi d^2}{4}} = (5.56\ldots)r^3$

In Bhāskara's *Līlāvatī* (see [3]) the volume is given as $= \frac{4\pi r^2 \times 2r}{6} = \frac{4\pi r^3}{3}$

Śrīdhara gives volume of a sphere $= \frac{38}{9}d^3 = (4.22)r^3$.

6.10. Area of Convex and Concave Spherical Segment

In India, Mahāvīrācārya is the first mathematician to give the rule for finding the area of the curved surface of a spherical segment.

Figure 6.11.

The concave spherical surface is compared to *catvala* (pit or hole in the ground) and the convex spherical segment with *kurmanibha* (like the back of a tortoise)

परिधेश्च चतुर्भागो विष्कम्भगुणः स विद्धि गणितफलम् ।
चत्वाले कूर्मनिभे क्षेत्रे निम्नोन्नते तस्मात् ।।

paridheśca caturbhāgo viṣkambhaguṇaḥ sa viddhi gaṇitaphalam,
catvāle kūrmanibhe kṣetre nimnonnate tasmāt.

(7 *śl*25)

This means one-fourth of the circumference multiplied by the diameter gives the area. Similarly, in the case of concave and convex areas like that of sacrificial fire pit and like that of (the back of) the tortoise, (the required result is to be arrived at).

This means $A = \frac{\text{Circumference}}{4} \times$ *viskambha.*

Viṣkambha means extension, width and breadth in general and diameter in particular. So the above rule can be interpreted as

$$A = \frac{p}{4} \times \text{diameter} \quad \text{where } p = \text{circumference} = \pi c$$

चत्वालक्षेत्रस्य व्यासस्तु भसङ्‍ख्यकः परिधिः ।
षट्पञ्चाशद्दृष्टं गणितं तस्यैव किं भवति ।।

catvālakṣetrasya vyāsastu bhasaṅkhyakaḥ paridhiḥ,
ṣaṭpañcāśaddṛṣṭaṁgaṇitaṁ tasyaiva kiṁ bhavati.

(7 *śl*26)

In the case of the area of a sacrificial fire-pit the measure of the diameter is 27 and the measure of the circumference is seen to be 56. What is the calculated measure of the area of that same (pit)?

Given diameter (*Viṣkambha*) = 27 and circumference = 56

$$A = \frac{\text{Circumference}}{4} \times viskambha = \frac{27}{4} \times 56 = 378$$

6.11. Ring

The calculation of the area of an annulus is a peculiarity of *jaina* texts. Mahāvīrācārya gives the formulae of the area of an out-lying and in-laying annular figure:

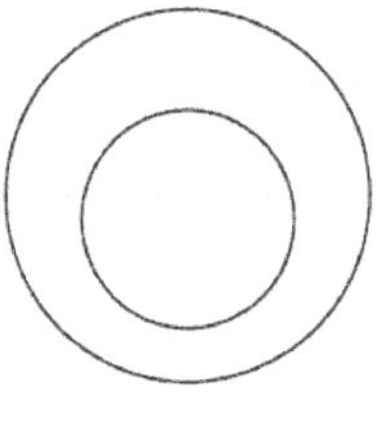

Figure 6.12.

निर्गमसहितो व्यासस्त्रिगुणो निर्गमगुणो बहिर्गणितम् ।
रहिताधिगमव्यासाभ्यन्तरचक्रवालवृत्तस्य ।।

nirgamasahito vyāsastriguṇo nirgamaguṇo bahirgaṇitaṁ
rahitādhigamavyāsādabhyantaracakravālavṛttasya.

(7 *śl*28)

This means the (inner) diameter, to which the breadth (of the annulus) is added, is multiplied by 3 and by the breadth (of the annulus). This gives rise to the value of the area of the outreaching annulus. Similarly, the measure of the in-lying annular figure is to be obtained from the diameter as diminished by the breadth of the annular area.

Then area $= (d \pm t)\sqrt{10}t$, where d is the inner diameter and t the width of the annulus.

Nārāyaṇa Paṇḍita also gives the area of a wheel like figure, "Add the square of the width to the product of the width and the diameter of the inner circle. Thrice the sum happens to be the area of a wheel like figure".

Let D and d be the diameter of the outer and the inner circles of a wheel, respectively and w be its width. The rule states that (gross) area of the wheel $= 3(dw + w^2)$, which follows by taking $\pi = 3$, and $D = d + 2w$.

6.12. Conclusion

In this paper we have presented Mahāvīrācārya's discussion of the areas and circumferences of the usual geometrical figures as well as some special curves. We have pointed out that Mahāvīrācārya is unique among Indian mathematicians in studying these special curves including the *āyatavṛtta* (ellipse). We have also compared Mahāvīrācārya's results with some other well-known Indian contributors.

Acknowledgements

The authors would like to thank the referees for their valuable comments and suggestions.

References

[1] *Gaṇita-Sāra-Sāṅgraha* of Mahāvīrācārya-
(1) edited with English translation by M. Rangacharya, Madras, 1912;
(2) Hindi translation by L.C. Jain, Shlapur, 1963;

(3) Kannada translation by Padmavtamma, Sri Hombuja Jain Math, Shimoga, 2000.

[2] Datta B.B., and Singh A.N. *History of Hindu Mathematics* (2 parts), Reprinted, Asia Publishing House, Bombay, 1962; Reprinted by Bharatiya Kala Prakashan, Delhi, 2001.

[3] Datta B.B., and Singh A.N. *History of Hindu Geometry*, Revised by K.S. Shukla, IJHS, 15(2), New Delhi, 1980.

[4] Saraswathi Amma, T.A. *Geometry in ancient and Medieval India*, MotilalBanarsidass, Delhi, 1979.

https://doi.org/10.1142/9789811255502_0007

Article 7

Kuṭṭīkāra Equations in Mahāvīrācārya's *Gaṇitasārasaṅgraha*

Sudarshan B

Department of Electronics and Telecommunication
RV College of Engineering
Bengaluru 560059, India
sudarshan.balaji2000@gmail.com

N. Shivakumar

Department of Mathematics
RV College of Engineering
Bengaluru 560059, India
shivakumarnswamy56@gmail.com

Mahāvīrācārya was a pioneering mathematician. His work Gaṇitasārasaṅgraha gives, in particular, an analysis of Indeterminate equations of the first order, also called as **Linear Diophantine Equations**, which he preferentially named "Kuṭṭīkāra". Mahāvīrācārya has discussed Linear Indeterminate equations of the form: $Ax - By = C$, under *five* different streams. They are: (a) Vallikā Kuṭṭīkāra (b) Viṣama Kuṭṭīkāra (c) Sakala Kuṭṭīkāra (d) Vicitra Kuṭṭīkāra (e) Suvarṇa Kuṭṭīkāra. In this paper, we give an exposition of these methods along with Āryabhaṭa's original motivation for Kuṭṭīkā (or Kuṭṭīkāra).

Keywords: Kuṭṭīkāra, Gaṇitasārasaṅgraha, Mahāvīrācārya, Vallikā, Viṣama, Sakala, Suvarṇa, Vicitra.

Mathematics Subject Classification 2020: 01-06, 01A32, 01A99

7.1. The Kuṭṭaka: An Introduction

Mahāvīrācārya (or *Mahāvīra*) was a Digambara Jain who lived in the later part of the rule of *Amoghavarṣa Nṛpatuṅga* (815–877 CE), a great and benevolent king of the *Raṣṭrakūṭa* dynasty, mainly ruling over north Karnataka and also other parts of India [1, 6]. The celebrated Jaina mathematician, *Mahāvīra*, wrote the *Gaṇitasārasaṅgraha* (GSS) and the uniqueness of this work is that, it was being used as a textbook for a very long time, especially in South India. *Mahāvīra* was also familiar with the works by his predecessors like *Āryabhaṭa*, *Bhāskara* (the first) and also *Brahmagupta* [1]. He has not only provided his own examples, but also has given detailed and improvised examples of the topics discussed by his predecessors. The text of *GSS* contains topics in arithmetic, algebra, geometry, mensuration, calculations regarding excavations and shadows spread over 9 chapters with about 1100 ślokas.

Indeterminate equations were systematically studied for determination of integral solutions first by ancient mathematicians in India. While Diophantus was content with finding one rational solution, Indians made an in-depth study by investigating all integral solutions of Indeterminate equations of the first and second order. By the 5th century CE, ancient Indians had discovered a general method for solving the first order Indeterminate equations in two variables. In this paper, we discuss the methods in the area of Indeterminate equations described by *Mahāvīrācārya* in his *Gaṇitasārasaṅgraha.*

7.2. Kuṭṭaka and its Types

Kuṭṭa, Kuṭṭaka, Kuṭṭakạra and *Kuṭṭīkāra*, the names used for the process of solving indeterminate equations, are all Sanskrit words derived from the root "*Kuṭṭ*", which means 'to grind' or 'pulverize'. *Mahāvīrācārya* had preferentially used the word "*Kuṭṭīkāra*" [4, 5].

Mahāvīrācārya, in his *Gaṇitasārasaṅgraha*, has discussed the analysis of Indeterminate equations of the first order. The analysis involves five methods. They are called: (a) *Vallikā Kuṭṭīkāra* (b) *Viṣama Kuṭṭīkāra* (c) *Sakala Kuṭṭīkāra* (d) *Suvarṇa Kuṭṭīkāra* and (e) *Vicitra Kuṭṭīkāra*.

Of these the last two may be considered as auxiliary methods or methods involving application of *Kuṭṭaka*; in this paper, the terms '*Kuṭṭaka*' and '*Kuṭṭīkāra*' are treated as synonymous. In general, the main objective of *Kuṭṭīkāra* is to find, given positive integers A, B, and C, the least positive integral multiple Ax of A such that when C is added to, or subtracted from, Ax, the result is exactly divisible by B, giving an integer y. Algebraically this can be represented as:

$$Ax - By = C \tag{1}$$

where A and B are positive integers and C is any integer. For reference, in the sequel, we mention that the term C above is referred to as the *mati* of the problem/equation.

7.2.1. *Vallikā Kuṭṭīkāra*

The rule underlying the process of calculation known as *Vallikā* in relation to *Kuṭṭīkāra* (which is a special kind of division or distribution), with reference to *śloka* 115 1/2, is as follows:

छित्वा छेदेन राशिं प्रथमफलमपोह्याप्तमन्योन्यभक्तं
स्थाप्य ऊर्ध्वाधः यतोऽधो मतिगुणमयुजाल्पे अवशिष्टे धनर्णम्।
छित्वाधः स्वोपरिघ्नोपरियुतहरभागाधिकाग्रस्य हारं
छित्वा छेदेन साग्रान्तरफलं अधिकाग्रान्वितं हारघातम्॥

The process explicated in this may be described in 4 major steps as follows:

Step (1): Mutual division i.e., dividing the dividend by the divisor to obtain the series of quotients.

$$
\begin{array}{lllllll}
A) & B & (q_1 & \rightarrow \text{neglected} & & & \\
 & \ldots & & & & & \\
\hline
 & r_1) & A & (q_2 & & & \\
 & & \ldots & & & & \\
\hline
 & & r_2) & r_1 & (q_3 & & \\
 & & & \ldots & & & \\
\hline
 & & & r_3) & r_2 & (q_4 & \\
 & & & & \ldots & & \\
\hline
 & & & & r_4) & r_3 & (q_5 \\
 & & & & & \ldots & \\
\hline
 & & & & & r_5 & \\
\hline
\end{array}
$$

Step (2): The first quotient q_1 is neglected and the other quotients are written down on below the other. An optionally chosen number (usually 1) is chosen. To this number, the value of the *mati* is added or subtracted (depending on the sign of the *mati*). This sum or difference and the optionally chosen number are written below the chain which has been obtained from the previous process. It is to be noted that the (optionally) chosen number should be such that its sum/difference with the *mati* must be divisible by the divisor pertaining to the last remainder, i.e., r_4 in this generalization.

Thus, the chain is fully formed. The final form of the chain would look something as shown below: If the optionally chosen number is represented as 'p', then,

$$
\begin{array}{c}
q_2 \\
q_3 \\
q_4 \\
q_5 \\
p \\
C + p
\end{array}
$$

This totally constitutes the formation of quotient chain process.

Step (3): Depending on the quotient-chain obtained in the previous step, we perform a set of multiplications and additions to obtain the group — value. Given that the value of the *mati* is 'C' and the optionally chosen number is 'p', the procedure to find the group — value is as shown:

The penultimate number (viz. p) is multiplied with the number above it (viz. q_5) and to this product, the number below it (viz. Q) is added. This new sum is now written in place of q_5 and a new chain is written down again with this new value of q_5. This process of multiplications and additions continue till the new value for the top-most element of the chain is found. This new value (of the top-most element) becomes the "Group-Value".

q_2	q_2	q_2	q_2	$((q_2 \times p_1) + p_2) \rightarrow G$
q_3	q_3	q_3	$((p_2 \times q_3) + p_3) \rightarrow p_1$	p_1
q_4	q_4	$((p_3 \times q_4) + p) \rightarrow p_2$	p_2	
q_5	$((p \times q_5) + Q) \rightarrow p_3$	p_3		
p	p			
$C + p = Q$				

Step (4): The final step is to divide the obtained group value 'G' by the divisor of the given problem (viz. 'B'). The remainder obtained from this division is the required *least integral solution* for the value of 'x', which is as shown below.

$$
\begin{array}{r}
B)\quad G \quad (q \\
\ldots \\
\hline
x \\
\hline
\end{array}
$$

The value of 'y' can be found by substituting the obtained value of 'x' in the equation. This completes the process. For better understanding, an example is given below.

Example: Let us consider the following with reference to ślokas 116 1/2–117 1/2

दृष्टास्सप्तत्रिंशत् कपित्थफलराशयो वने पथिकैः।
सप्तदशापोह्य हृते व्येकाशीत्यांशकप्रमाणं किम् ।।

"In the bright and refreshing outskirts of a forest, full of numerous trees with branches bent down with the weight of flowers and fruits, trees such as jambu trees, lime trees, plantain, areca palms, jack trees, date palms, hintala trees, palyras, punnaga trees and mango trees- the various quarters of which were filled with the many sounds of crowds of parrots and cuckoos near found springs containing lotuses with bees swarming around them — a number of weary travelers entered with joy."

"There were 63 equal heaps of plantain fruits put together with 7 more of the same fruits. These were distributed equally among 23 travelers so as not to leave a remainder. Tell me now the number of fruits in each heap."

***Solution*:** Algebraically representing the given problem, we need to: Determine x such that the expression $\frac{63x+7}{23}$ turns out to be a whole number/positive integer.

Given:

$$y = \frac{63x + 7}{23} \tag{2}$$

Step (1): We start by dividing the *rāśi* or the divisor, i.e., 23 in the given problem, by the 'dividend', i.e., 63 in the given problem, and continue the process of division, just as finding the h.c.f or g.c.d of the numbers, as shown below.

23)	63	(2	→ neglected			
	−46					
	17)	23	(1			
		−17				
		6)	17	(2		
			−12			
			5)	6	(1	
				−5		
				1)	5	(4
					−4	
					1	

It is important to note that we stop the division with the fifth remainder. This is so because, for this method to work, the least remainder (i.e., 1) is to be obtained in the odd position of order in the series of divisions that has been done. Also, it is to be noted that the remainder should never be made zero, and the remainder obtained, in the odd position of order in the series of divisions made, must be the least possible.

Step (2): Neglecting the first quotient (i.e., 2), the rest of the quotients so obtained are written one below the other as shown. This looks like a chain and hence the name *Vallikā*.

Now, we choose a number (usually 1), such that, when multiplied by the last remainder, i.e., 1 and then later combined with the *mati*, i.e. 7, would be divisible by the divisor pertaining to the last remainder (viz., 1). In other words, the number obtained after this process is a whole number. The number chosen and the sum obtained are both written below the quotient chain. Note that the number to be combined changes with given data.

$$\begin{array}{c} 1 \\ 2 \\ 1 \\ 4 \\ 1 \\ 8 \end{array}$$

Step (3): We note that the penultimate number in this chain is 1, and the last number is 8. According to the rule, the penultimate number is multiplied by the number above it and this product is added with the number present below the penultimate number. This means, 1 is multiplied by 4 (the number above) and to this product (viz., 4), the number below (i.e., 8) is added resulting in 12. Now re-write the creeper chain replacing 4 with 12 and neglecting the last number (i.e., 8 in this case).

Now, in the new table formed, 12 is the penultimate number and 1 is the last number. Repeat this procedure of multiplication and addition till the '*group value*' is found.

1	1	1	1	$((38 \times 1) + 13) \to 51$
2	2	2	$((13 \times 2) + 12) \to 38$	38
1	1	$((12 \times 1) + 1) \to 13$	13	
4	$((1 \times 4) + 8) \to 12$	12		
1	1			
8				

The group — value obtained is 51.

Step (4): Now, this group value found should be divided by the *rāśi* or divisor of the given problem, and the remainder so obtained by this division is our required solution for the number of heaps. This is as shown below.

```
23)  51  (2
    -46
    ----
      5
    ----
```

$$\text{Hence, } x = 5; \quad y = \frac{63 \times 5 + 7}{23} = 14.$$

Procedure for solving 2 simultaneous Indeterminate equations:

Following the exposition of the *Vallikā Kuṭṭīkāra* discussed above, GSS includes various examples, in the form of exercises, involving 2 or more simultaneous equations as in (1), with one variable common to all equations and the other varying with each equation. While there is no verse in GSS explicitly describing the procedure for this, the following method seems to have been meant to be used in solving the exercises. For convenience we describe the steps here in the case of 2 equations; it may be seen that it can be readily adopted when there are more equations.

Step (1): As a first step, we find the group-values of both the given equations separately, with their respective *mati.*

Step (2): The divisor related to the larger group-value is divided by the divisor related to the smaller group-value, so as to form the new *Vallikā*-chain.

Step (3): The group value of this new *Vallikā*-chain is calculated and it is later divided by the divisor of the smaller group-value (As for the optionally chosen number, we usually choose 1, and the new *mati* will be the difference of the two group-values obtained).

Step (4): The remainder thus obtained is multiplied by the divisor related to the larger group-value and this product is added to the larger group value. This sum yields the value of 'x' we require.

Example: Let us consider the following with reference to *ślokas* 127 1/2:

दृष्टा जम्बूफलानी पथि पथिकजनै राशयस्तत्र राशी द्वौ त्र्यग्रौ तौ नवानां त्रय इति पुनरेकादशानां विभक्ताः ।
पञ्चाग्रास्ते यतीनां चतुरधिकतराः पञ्च ते सप्तकानां कुट्टीकारार्थविन्मे कथय गणकसञ्चिन्त्य राशिप्रमाणं ॥ १२७ १/२ ॥

127 1/2: "The travelers saw on the way, certain (equal) heaps of *jambū* fruits. Of them 2 (heaps) were equally divided among 9 ascetics and left 3 (fruits) as remainder. Again 3 (heaps) were similarly divided among 11 persons and the remainder was 5 fruits; then again 5 of those heaps were similarly divided among 7 and there were 4 more fruits (left out) of them. O you Arithmetician who know the meaning of the *Kuṭṭīkāra* process of distribution, tell me after thinking out well the numerical measure of a heap (here)."

***Solution*:** This may be dealt with in terms of three equations $2x - 3 = 9y$, $3x - 5 = 11z$ and $5x - 4 = 7w$ where x is the size of the heap and y, z and w are the number of travelers in the 3 instances. It may be noted that these equations can be further simplified into two equations by considering:

$$\frac{9y+3}{2} = \frac{11z+5}{3} \tag{3}$$

and

$$\frac{11z+5}{3} = \frac{7w+4}{5}. \tag{4}$$

The above equations may be solved by following the steps as described above, resulting in a value of x satisfying all the equations.

7.2.2. *Viṣama Kuṭṭīkāra*

Here is the verse from Gaṇitasārasaṅgraha (with reference to *śloka* 134 1/2) explaining the process of *Viṣama Kuṭṭīkāra.*

विषमकुट्टीकारस्य सूत्रम् -

मतिसङ्गुणितौ छेदौ योज्योनत्याज्यसंयुतौ राशिहृतौ ।
भिन्ने कुट्टीकारे गुणकारोऽयं समुद्दिष्टः ॥ १३४ १/२ ॥

There seem to be uncertainties about the interpretation of this verse in literature. We shall therefore not include any details here. It is seen however that this applies equations of type (1) which do not admit solutions in integers; the verse is about rational equations. The procedure does not involve mutual division, that was crucial to *Vallikā Kuṭṭīkāra*, but nevertheless seems to be included in *Kuṭṭīkāra*, by extension. As an exercise on the process, Mahāvīrācārya includes the following example with reference to *śloka* 135 1/2 given below.

राशिः षट्के न हतो दशान्वितो नवहृतो निरवशेषः ।
दशभिर्हीनश्च तथा तद्गुणको कौ ममाशु सङ्कथय ॥

This asks for x for which $6x + 10$ is divisible by 9 (similarly with $6x - 10$), which is possible only in rationals, and not in integers.

7.2.3. *Sakala Kuṭṭīkāra*

Motivation for the method: When the value of the *mati* is huge, usage of the *Vallikā Kuṭṭīkāra* method became very tedious and a laborious process. *Sakala Kuṭṭīkāra* introduces a technique to facilitate dealing with this issue. This method turned out to be beneficial in the applications of *Kuṭṭīkāra* in the field of Astronomy, where high values of x, y, and *mati* are encountered. Before we proceed with the example problem, we should note that, if there are any common factors that can be taken out from the numerator or denominator, it should be taken out. The method should be applied to the equation in its basic form or simplified form.

The method: The rule in relation to this is:

सकलकुट्टीकारस्य सूत्रम् -

भाज्यच्छेदाग्रशेषैः प्रथमहृतिफलं त्याज्यमन्योन्यभक्तं न्यस्यन्ते साग्रमूर्वँरुपरिगुणयुतं तैस्समानासमाने ।
स्वर्णघ्नं व्याप्तहारौ गुनधनमृणयोश्चाधिकाग्रस्य हारं हृत्वा हृत्वा तु साग्रान्तरधनमधिकाग्रान्वितं हारघातम् ॥ १३६ १/२ ॥

The procedure, with reference to *śloka* 136 1/2, is very similar to the *Vallikā Kuṭṭīkāra* method except for 2 instances — the calculation of *mati* and the values of the solution. For better understanding of the method, an example, with reference to *śloka* 137 1/2, is included.

Example: Consider an example with reference to *śloka* 137 1/2

सप्तोत्तरसप्तत्या युतं शतं योज्यमानमष्टत्रिंशत् ।
सैकशतद्वयभक्तं को गुणकारो भवेदत्र ॥

137 1/2: "One hundred and seventy-seven (is the dividend-co-efficient of the unknown factor), 240 is the known quantity associated (with the product so as to be added to or subtracted from it); the whole is divided by 201 (and leaves no remainder). What is the (unknown) factor here (with which the given dividend — coefficient is to be multiplied)?"

***Solution*:** In the modern notation the problem corresponds to the following, where the second expression is obtained by canceling the common factor.

$$\frac{177x \pm 240}{201} = \frac{59x \pm 80}{67} \tag{5}$$

The procedure of Sakala is almost the same as the first method. The last two numbers in the chain are to be found in a different way.

The quotient chain is written after mutual division of the dividend by the divisor. Now, for the last two numbers, an optionally chosen number, say 1, and its sum with the *mati* are to be taken. Now, the *mati* is calculated in a different way. The *agra*[1] in the given problem (viz., 80) is divided by the divisor of the problem (viz., 67). The remainder of this is the new *mati* which turns out to be 13. The chain is therefore:

$$\begin{array}{c} 1 \\ 7 \\ 2 \\ 1 \\ 1 \\ 1 \\ 14 \end{array}$$

As before, we now calculate the group value of this chain which turns out to be 392. Now, this group — value is divided by the divisor and the remainder is one of the solutions of x.

The value of $x = 57$ (the remainder obtained) and the other solution of x is the difference of the divisor and the remainder, i.e., $67 - 57 = 10$.

Now, in order to determine which solution corresponds to which equation, we need to find out the length of the chain. If the number of elements in the chain is an even number, the remainder obtained corresponds to the equation with the positive *agra*, and the other solution corresponds to the other equation, and vice versa, when the number of elements in the chain is an odd number. Let us verify this by substituting the values of x.

$$y_1 = \frac{59x + 80}{67} = \frac{59(10) + 80}{67} = 10 \tag{6}$$

$$y_2 = \frac{59x - 80}{67} = \frac{59(57) - 80}{67} = 49. \tag{7}$$

[1] *agra* and *mati* are one and the same.

Since, the obtained answers are whole numbers, the answers obtained are correct and the values obtained are the values for y_1 and y_2.

7.2.4. *Suvarṇa and Vicitra Kuṭṭīkāra*

The Suvarṇa *Kuṭṭaka* is a kind of auxiliary method and essentially concerns calculations involving gold of various purities, finding the purity of a mixture and so on. There is no general rule for this *Kuṭṭaka*, instead, this *Kuṭṭaka* is a collection of rules. This bifurcation may have been done in order to ease the readers (or people who used this work as a textbook for reference during those times) in understanding the mathematical concepts involved and also the practical usage of the concepts, since, making mixed gold, finding the weight of gold present in a specific varṇa,[2] etc., are all required components to be known during the forging process.

Similarly, the *Vicitra Kuṭṭīkāra*, like the previous *Kuṭṭaka*, is an auxiliary method. It does not have any specific general rule, but is a collection of rules, considering different scenarios. *Vicitra* means unusual or peculiar, in Sanskrit (as also in Kannada). Some of rules provided under this *Kuṭṭaka* include combinatorics, truthful and untruthful statements, arrangement of arrows, problems involving money/capital, etc. These may be treated as methods where we find the application of the concept of kuṭṭaka that were prevalent during the ancient times.

7.3. Āryabhaṭa's Approach and Kuṭṭaka

We conclude with a brief history of the Kuṭṭaka method. A systematic approach to find solutions of linear indeterminate equations was started by Āryabhaṭa (499 CE) and further work was carried by different mathematicians like Brahmagupta, Mahavīra etc. Āryabhaṭa gives the principle behind the Kuṭṭaka in just 2 verses (verses 32 and 33 under the section *Gaṇita*) [2, 3]. Bhāskara I played a vital

[2]varṇa refers to the proportion of pure gold present in any given piece of gold.

role in elaborating the meaning of these verses and he has also given about 24 examples relating to Kuṭṭaka, applied in the field of Astronomy.

Āryabhaṭa's problem was to find a number (N), such that, when N was divided by two divisors (a, b), they yielded remainders (R_1, R_2). Mathematically,

$$y = ax + R_1 = by + R_2 \tag{8}$$

which readily reduces to an equation of the form $Ax + C = By$ considered here. The basic principles of the Kuṭṭaka method were introduced by Āryabhaṭa in this context. Subsequently there were various improvements in the handling of the process, organizing the sequence of divisors, and other finer detail, including in the work of Mahāvīrācārya.

Acknowledgment

First of all, the first author expresses his thanks to his parents for their consistent support. Secondly, both the authors are thankful and express their gratitude to Dr. Balachandra Rao, Hon. Director of Bhavan's Gandhi Centre of Science and Human Values, Bengaluru (India), for his timely help and support towards this paper. The authors are also thankful to Dr. Anupam Jain, Indore (India), who has also lent his support for us. Finally, the authors are thankful to the referee for scrutinizing the details in the paper and also for their valuable suggestions and comments.

References

[1] Dr. S. Balachandra Rao, Indian Mathematics and Astronomy: Some Landmarks, 2000.

[2] Bibhutibhushan Datta and Avadhesh Narayan Singh, "A History of Hindu Mathematics: A Source Book" Part I and II, 1962.

[3] A. K. Dutta, Mathematics in ancient India., Reson 7, 6–22 (2002). https://doi.org/10.1007/BF02835539.

[4] Dr. (Mrs). Padmavathamma, The Gaṇitasārasaṅgraha of Mahāvīrācārya, with Kannada Translation and Notes, 2000.

[5] Prof. Rangācārya, The Gaṇitasārasaṅgraha of Mahāvīrācārya, with English Translation and Notes, Madras Govt. Press, Madras 1912.
[6] Prof. M. S. Sriram, — Nptel.ac.in. NPTEL: "Mathematics in India — From Vedic Period to Modern Times" — Lecture Module — 16, GSS.

https://doi.org/10.1142/9789811255502_0008

Article 8

Series Summations in Śrīdhara and Mahāvīra's Works

R. C. Gupta

R-20, Ras Bahar Colony, P.O. Sipri Bazar
Jhansi 284003, U.P., India

The arithmetical progression (A.P.) has been known in all cultural areas since ancient times. The natural series

$$1 + 2 + 3 + \cdots \qquad (*)$$

formed from positive integers or those formed from the squares or cubes of terms of (*) were also known in ancient Indian mathematics including the Jaina School. Triangular numbers are formed by the sums of (*) upto $1, 2, 3, \ldots$ terms respectively. In Śrīdhara (8th century CE) we meet the idea of forming a series whose terms are the successive sums of (*) upto b, $(b+e)$, $(b+2e), \ldots$ terms respectively, where b and e are positive integers. Mahāvīrācārya gave correctly an explicit rule for finding the sum of n terms of such a series. In this paper we recall the results of Śrīdhara and Mahāvīra on the topic and put the theme involved in a broader perspective. Some lapses on the part of the earlier modern scholars are pointed out.

Keywords: Arithmetical progression, polygonal numbers, compound series, Śrīdhara, Mahāvīrācārya, ancient Indian mathematics.

Mathematics Subject Classification 2020: 01-06, 01A32, 01A99.

8.1. Introduction

In this paper the discussion is confined to series of finite number of terms. We are mostly concerned with Arithmetical Progressions

(A.P.) and some other series based on or derived from them. For an A.P. the general term is

$$T_r = a + (r - 1)d \tag{1.1}$$

where a is the first term (*ādi*, *mukha*, etc.) and d is the common difference (*uttara*, *caya*, etc.). In the simplest A.P., the terms are the natural (whole) numbers or positive integers:

$$1 + 2 + 3 + 4 + \cdots + r + \cdots . \tag{1.2}$$

For the general A.P. namely,

$$a + (a + d) + (a + 2d) + (a + 3d) + \cdots \tag{1.3}$$

the sum to n terms is,

$$S_n = \left(\frac{n}{2}\right)[2a + (n - 1)d] = \left(\frac{1}{2}\right)[(1^{st}\,term) + (last\,term)] \cdot n \tag{1.4}$$

Some historical and educational aspects of A.P. have been dealt by the present author [1].

Two more series are found in ancient mathematical works:

$$1^2 + 2^2 + 3^2 + \cdots + n^2 \tag{1.5}$$

$$1^3 + 2^3 + 3^3 + \cdots + n^3. \tag{1.6}$$

In some ancient Indian works the rules for the sum of the following two series are also found:

$$a^2 + (a + d)^2 + (a + 2d)^2 + \cdots\ to\, n\ terms \tag{1.7}$$

$$a^3 + (a + d)^3 + (a + 2d)^3 + \cdots\ to\, n\ terms. \tag{1.8}$$

Due to ancient concept of a three-dimensional space, discussion of rules was mostly confined to linear (*sūcya*), areal (*pratara*) and spatial (*ghana*) cases. Chauthaiwale [2] contains a survey of various formulas for sum of the series (mentioned above) as given by eminent ancient mathematicians of India. The survey shows that although rules for summing (1.5) and (1.6) were already known to Āryabhaṭa (5th cent. CE) and Brahmagupta (early 7th cent.), those for (1.7) and (1.8) are found given by subsequent authors such as Śrīdhara (middle 8th cent.) and Mahāvīra (9th cent.).

For a fine method of finding the sum of

$$a^k + (a+d)^k + (a+2d)^k + \cdots \textit{ to } n \textit{ terms} \tag{1.9}$$

where k is a positive integer, a recent paper by Imam [3] may be consulted (it gives also a brief history of earlier methods).

Now consider any given series

$$w_1 + w_2 + w_3 + w_4 + \cdots . \tag{1.10}$$

Let t_r be the sum (*yoga*) of the first r terms of (1.10) i.e.,

$$t_r = w_1 + w_2 + w_3 + \cdots + w_r. \tag{1.11}$$

Then the series

$$t_1 + t_2 + t_3 + \cdots \tag{1.12}$$

will be called the *yoga* series of the series (1.10). Thus, the *yoga* series of the natural series (1.2) will be,

$$1 + (1+2) + (1+2+3) + (1+2+3+4) + \cdots \tag{1.13}$$

or,

$$1 + 3 + 6 + 10 + \cdots + r\frac{(r+1)}{2} + \cdots . \tag{1.14}$$

Interestingly the terms of this last series can be represented by diagrams as follows:

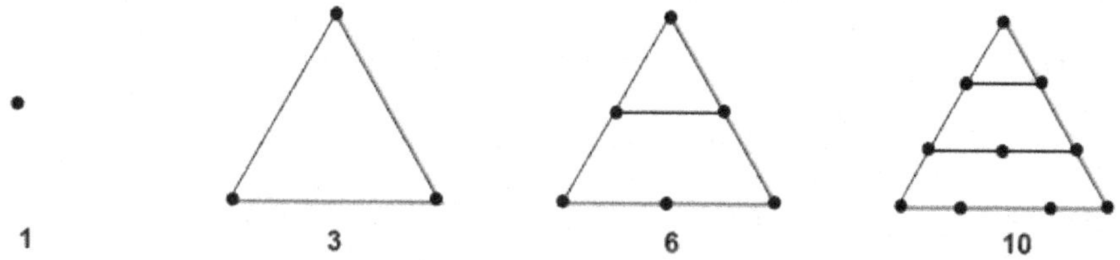

Consequently, the numbers

$$n\frac{(n+1)}{2}, \quad n = 1, 2, 3, \ldots \tag{1.15}$$

are called triangular numbers. Similar illustration may be given for the square numbers of the series,

$$1 + 4 + 9 + 16 + \cdots + r^2 + \cdots \tag{1.16}$$

and for the pentagonal numbers given by,

$$n\frac{(3n-1)}{2}, \quad n = 1, 2, 3, \ldots. \tag{1.17}$$

The general polygonal numbers of order g (correspond to polygon of order g) are given by [4],

$$P(n) = \left(\frac{n}{2}\right)[(g-2)(n-1)+2], \quad \textit{where } n = 1, 2, 3, 4, \ldots. \tag{1.18}$$

For a detailed exposition of the polygonal or figurate numbers in general from the Jaina sources a paper by Jadhav [5] may be consulted.

8.2. Formation of Peculiar Series

It may be noted that the *yoga* series (1.13) is formed from the natural series (1.2) by taking latter's sum of $1, 2, 3, 4, \ldots$ terms successively. Instead of sum of $1, 2, 3, \ldots$ terms, Śrīdhara formed a (super) *yoga* series by taking sum of b, $(b + e)$, $(b + 2e), \ldots$ terms of (1.3) successively. Also, we consider such formation of new series now from any A.P. instead of just (1.2). We give the details.

Let there be a given A.P. (of sufficient number of terms) as,

$$a + (a + d) + (a + 2d) + \cdots + [a + (r-1)d] + \cdots \tag{2.1}$$

or, say,

$$u_1 + u_2 + u_3 + \cdots + u_r + \cdots. \tag{2.2}$$

Also let there be also any chosen (*iṣta*) A.P. of n terms:

$$b + (b + e) + (b + 2e) + \cdots + [b + (r-1)e] + \cdots + [b + (n-1)e] \tag{2.3}$$

or say,

$$v_1 + v_2 + v_3 + \cdots + v_r + \cdots + v_n. \tag{2.4}$$

Suppose the new type or super-*yoga* series to be formed is,

$$T_1 + T_2 + T_3 + \cdots + T_r + \cdots + T_n. \tag{2.5}$$

These terms are formed as follows:

$$1^{\text{st}} \text{ term, } T_1 = u_1 + u_2 + u_3 + \cdots \text{to } b \text{ terms i.e., upto } u_b.$$

$$2^{\text{nd}} \text{ term, } T_2 = u_1 + u_2 + u_3 + \cdots, \text{to } (b+e) \text{ terms i.e., upto } u_{b+e} \tag{2.6}$$

and so on (it is clear that b and e should be +ve integers).

$$r^{\text{th}} \text{ term, } T_r = u_1 + u_2 + u_3 + \cdots \text{to}[b + (r-1)e] \text{ terms i.e., upto } u_{b+(r-1)e} \tag{2.7}$$

....

$$n^{\text{th}} \text{ or last term } T_n = u_1 + u_2 + u_3 + \cdots \text{to } n \text{ terms i.e., upto } u_{[b+(n-1)e]}. \tag{2.8}$$

Thus, we see that T_1 has v_1 number of terms, T_2 has v_2 number of terms and so on such that T_r has v_r number of terms. But the point to note is that *each* T_r is formed from the terms of the given A.P. (2.1) starting with the first term a every time. We concentrate on T_r. It has $[b + (r-1)e]$ terms of the u-series (2.2) or the series (2.1). In this way,

$$\text{the first term in } T_r, \text{ i.e., in (2.7), is } u_1 = a, \tag{2.9}$$

and the *last* term in T_r is the $[b + (r-1)e]^{\text{th}}$ term of (2.1).

That is, the last term u_l of T_r is given by, using (1.1),

$$u_l = a + [\{b + (r-1)e\} - 1]d \tag{2.10}$$

Finally, we recollect the rule

$$\text{sum of A.P.} = \left(\frac{1}{2}\right)(u_1 + u_l)\,(no.\,of\,terms) \tag{2.11}$$

In this way we get

$$T_r = \left(\frac{1}{2}\right) \cdot [2a + (b-1)d + (r-1)de] \cdot [b + (r-1)e] \quad (2.12)$$

$$= \left(\frac{1}{2}\right) \cdot [k + (r-1)de] \cdot [b + (r-1)e] \quad (2.13)$$

where,

$$k = 2a + (b-1)d. \quad (2.14)$$

Thus, the sum of n terms of the new series (2.5) will be,

$$S_n = \sum_{r=1}^{n} T_r = \left(\frac{1}{2}\right) \sum_{r=1}^{n} [k + (r-1)de][b + (r-1)e] \quad (2.15)$$

which leads to,

$$2S_n = \sum_{r=1}^{n} [kb + (k + bd)e(r-1) + de^2(r-1)^2] \quad (2.16)$$

On summing each term in the square bracket, we get

$$2S_n = \left[kb.n + (k + bd)e \cdot \frac{(n-1)n}{2} + de^2 \cdot \frac{(n-1)n(2n-1)}{6}\right]. \quad (2.17)$$

Thus, we have

$$S_n = \left(\frac{n}{2}\right) \left[\frac{(n-1)(2n-1)}{6} de^2 + \frac{(n-1)}{2}(k + bd)e + kb\right]. \quad (2.18)$$

Śrīdhara [6] in his *Pāṭīgaṇita*, Rule 106 (text p.153 and transl. pp. 84–85) took the case of finding the sum of the series (2.5) when $a = 1$ and $d = 1$. That is, his Rule 106 is for finding,

$$[1 + 2 + 3 + \cdots t\ b\ \textit{terms}] + [1 + 2 + 3 + \cdots to (b + e)\ \textit{terms}]$$
$$+ [1 + 2 + 3 + \cdots to\ (b + 2e)\ \textit{terms}] + \cdots to\ n\ \textit{terms}. \quad (2.19)$$

Śrīdhara's result is equivalent in the modern form (transl.p.85)

$$S_n = \left(\frac{1}{2}\right)\left[\sum_{r=1}^{n}\{b+(r-1)e\}^2 + \sum_{r=1}^{n}\{b+(r-1)e\}\right]. \qquad (2.20)$$

This result is same as our (2.15) since (2.14) gives $k = (b+1)$ when $b = d = 1$. Although Śrīdhara had already given rules to find the sum of series involved in (2.20), no direct formula of his like (2.18) is known. However, he has added a numerical exercise on the topic (see next Section 3). Also note that number of terms in (2.1) is $[b+(n-1)e]$ at least.

8.3. The Śrīdhara-Mahāvīra Series

The great Jaina mathematician Mahāvīra (circa 850 A.D.) seems to be the first to give an explicit direct rule (in terms of b, e and n) to find the sum of the series (2.19). In his *Gaṇita-sāra-saṅgraha* (Rule VI, $305 - 305\frac{1}{2}$) he says [7]

द्विगुणैकोनपदोत्तरकृतिहतिरङ्गाहृता चयार्धयुता ।

आदिचयाहतियुक्ता व्येकपदघ्नादिगुणितेन ||३०५||

सैकप्रभवेन युता पददलगुणितैव चितिचितिका ||३०५$\frac{१}{२}$||

Dviguṇaikonapadottarakṛtihatiraṅgāhṛtā cayārdhayutā |
Ādicayāhatiyuktā vyekapadaghnādiguṇitena ||305||
Saikaprabhavena yutā padadalaguṇitaiva citicitikā $||305\frac{1}{2}||$

"Twice the number of terms (in the chosen series) is diminished by one and (then) multiplied by the square of the common difference. This product is divided by six and increased by half the common difference and (also) by the product of the first term and the common difference. The sum (so obtained) is multiplied by the number of terms as diminished by one and then increased by the product obtained by multiplying the first term as increased by one by the first term itself. The quantity (so resulting) when multiplied by half the number of terms (in the chosen series) gives rise to the required sum of the series wherein the terms themselves are sums (of natural series)."

That is,

$$S' = \left[\left\{\frac{(2n-1)e^2}{6} + \frac{e}{2} + be\right\}(n-1) + (b+1)b\right] \cdot \left(\frac{n}{2}\right). \quad (3.1)$$

Mahāvīra's formula can be easily seen to be the same (except for the arrangement of terms) as (2.18) when $a = d = 1$, and thus $k = (b+1)$ by (2.14).

We now take numerical problems on the series (2.19). Śrīdhara's Example 120 (text p. 153 and transl. p. 85) reads:

"O the best mathematician, say the sum of sums of the series of natural numbers (each beginning with 1), whose number of terms are the first six terms of the A.P. with 3 as the first term and 5 as the common difference."

The problem has been worked out by the ancient commentator (see text p.153) by two methods. He starts with the listing of the six terms of the chosen A.P. as

$$3,\ 8,\ 13,\ 18,\ 23,\ 28. \quad (3.2)$$

Then the six terms of the series (to be summed) are formed (by summing natural series upto above number of terms each time e.g., $1+2+3$ to 3 terms etc.) and listed as

$$6,\ 36\ (\textit{by taking sum to 8 terms}),\ 91,\ 171,\ 276,\ 406. \quad (3.3)$$

The sum of these six terms is 986 which is the required answer or 'sum of sums.'

The second method used by the commentator is called by him as *laghukarma* (short method). In this he applies Śrīdhara's rules to complete the two summations involved in (2.20) getting thereby $(1879 + 93)$. Half of this sum gives the required answer. Of course, if we apply Mahāvīra's direct formula (3.1) with the known $b = 3$, $e = 5$, $n = 6$, we get the expected same answer.

Mahāvīra's own example on the topic involves longer series of the type (2.19). In his *Gaṇita-sāra-saṇgraha*, Example $306\frac{1}{2}$ (text p. 105, transl. p. 173), the problem is to find the sum of the said series (2.19) when $b = 6$, $e = 5$, and $n = 18$.

This numerical example is given immediately after Mahāvīra's verbal rule implying the formula (3.1) which represents that rule in modern mathematical form. So, putting the given numerical values in (3.1) we get the required sum

$$S' = \left[\left\{\frac{35}{6} \times 25 + \frac{5}{2} + 30\right\} \times 17 + 42\right]\left(\frac{18}{2}\right) \tag{3.4}$$

$$= 27663, \textit{ on simplification.} \tag{3.5}$$

The same answer is found in Rangacharya's Appendix III, to GSS (p. 316).

Here the chosen series (A.P.) [cf. (3.2)] consists of the 18 terms:

$$6, 11, 16, 21, 26, \ldots, 86, 91. \tag{3.6}$$

The series, which was expected to be summed would be, by using (2.19) [cf. formation of (3.3) above],

$$21 + 66 + 136 + \cdots (\textit{total } 18 \textit{ terms}) \cdots + 4186. \tag{3.7}$$

It is clear that (3.7) is not an A.P. nor it was expected to be so (due to the manner in which its terms are formed). Thus Rangacharya (transl. of *GSS*, p. 172,f.n.) is wrong in stating that Mahāvīra's formula (3.1) "is the sum of the series in arithmetical progression, wherein each term of a series of natural numbers...". And other translators of *Gaṇita-sāra-saṇgraha* (such as L. C. Jain and Padmavathamma) who followed Rangacharya, repeated the mistake. Actually, it follows from (1.14) that each term of (3.7) will be a triangular number.

We have seen above that, Śrīdhara and Mahāvīra, both have played a role in the peculiar type of series discussed above. Series,

$$\begin{aligned}&[a + (a + d) + (a + 2d) + \cdots \text{to } b \textit{ terms}]\\&\quad + [a + (a + d) + (a + 2d) + \cdots \text{to } (b + e) \textit{ terms}] + \cdots\\&\quad + [a + (a + d) + (a + 2d) + \cdots \{\text{to } b + (n-1)e\} \textit{ terms}]\end{aligned} \tag{3.8}$$

may rightly be called Śrīdhara-Mahāvīra series. It is assumed that b and e are positive integers. The total number of terms in the Śrīdhara-Mahāvīra series (3.8) is still n which is the number of the terms in the chosen (*iṣṭa*) series, namely,

$$b + (b + e) + (b + 2e) + \cdots + [b + (n - 1)e]. \tag{3.9}$$

The peculiarity of the series (3.8) lies in the fact that *each term of it* represents the sum of the given A.P., namely,

$$a + (a + d) + (a + 2d) + \cdots \tag{3.10}$$

to a number (of terms) which itself is governed by the *chosen A.P. (3.9)'s particular term (serially) each time.* In other words, the various terms of (3.8) are same as denoted in (2.5) and explained fully thereafter in Section 2 above. So, the sum of the Śrīdhara-Mahāvīra series is S_n as given by (2.18). Its last (T_n) term is formed from the first $[b + (n - 1)e]$ terms of (2.1).

8.4. Epilogue and Concluding Remarks

The rth term of the Śrīdhara-Mahāvīra series is given by,

$$T_r = a + (a + d) + (a + 2d) + \cdots \textit{ to } [b + (r - 1)e] \textit{ terms.} \tag{4.1}$$

That is, T_r is a 'sum' of an A.P. So also is the case with all term T_1, $T_2, \ldots, T_n$ Since Śrīdhara-Mahāvīra series is the 'sum' of all these $T_1, T_2, T_3, \ldots$ it said to give or represent 'sum of sums'. So Śrīdhara called the sum of such series as *saṅkalita-saṅkalitam*, and Mahāvīra called the same as *citi-citikā* in his rule, but *citi-saṅkalita* in example.

By definition (1.12), the triangular series (1.14) is the *yoga* series of the natural series (1.2). The *yoga* series of the A.P. (3.9) upto n terms will be

$$b + (2b + e) + (3b + 3e) + (4b + 6e) + \cdots \textit{ to } n \textit{ terms.} \tag{4.2}$$

The general term of (4.2) will be

$$G_r = rb + r(r - 1)\frac{e}{2}. \tag{4.3}$$

Using this the sum of (4.2) can be easily found. The result is the sum

$$S = \left[\frac{n(n+1)}{2}\right] \cdot \left[b + (n-1)\frac{e}{3}\right]. \tag{4.4}$$

A. K. Bag [8] mentions (4.4) for the sum of (4.2). He also quotes equivalent of R.H.S. of (3.1) which is the modern form of Mahāvīra's Rule, VI, 305 – $305\frac{1}{2}$. But he wrongly states that Mahāvīra's said rule is for the series (4.2) and consequently he regards Mahāvīra's formula (3.1) as "an incorrect result". However, we have shown that Mahāvīra's said rule is *not* for the series (4.2) but is rather for the series (2.19). His formula (3.1) correctly gives the sum of the series (2.19) and (2.20).

References

[1] R. C. Gupta, "Historical and Expository Material for Teaching Arithmetical Progressions". *Indian Journal of Mathematics Education* (Delhi), Vol. 17, No. 3 (Jan. 1998), pp. 1–10.

[2] S. M. Chauthaiwale, "Indian Mathematicians on Sums of Terms in Arithmetical Progression". *Gaṇita Bhāratī* 27 (2005), 15–25.

[3] Faiz Imam, "Sum of rth powers of first n terms of an A.P." *Bulletin of the Kerala Mathematics Association*, 13(1) (June 2016) 51–56.

[4] J.K. Baumgart, D.E. Deal, B.R. Vogeli, A.E. Hallerberg, (editors), *Historical Topics for Mathematics Classroom.* N.C.T.M. 31st Yearbook, Washington D.C. 1969, p. 57 where credit for the formula (1.18) is given to Hypsicles (about 180 B.C.).

[5] Dipak Jadhav, "On the Figurate Numbers from the *Bhagavatī Sūtra.*" *Gaṇita Bhāratī* 31 (2009), 35–55.

[6] K. S. Shukla (editor), *The Pāṭīgaṇita of Śrīdharācārya* (with an ancient Sanskrit commentary) edited with English Translation and Notes. Dept. of Math., Lucknow Univ., 1959.

[7] M. Rangacarya (editor), *The Gaṇita-sāra-saṅgraha of Mahāvīrācārya*, edited with English Translation and Notes, Govt. of Madras, Madras (now Chennai), 1912 (In symbols, his a is our b, and his b is our e). The *GSS* was translated into Hindi by L. C. Jain (Sholapur, 1963). Padmavathamma included Rangacarya's translation in her edition of *GSS* (Hombuja, 2000). All page references to *GSS* in our paper are to Madras, 1912 work.

[8] A. K. Bag, *Mathematics in Ancient and Medieval India.* Chaukhambha Orientalia, Varanasi, 1979 (In symbols, his a is our b, and his b is our e), p. 185 (there are some misprints on the page).

https://doi.org/10.1142/9789811255502_0009

Article 9

Cryptography and Error-Detection in *Siribhūvalaya*

Anil Kumar Jain*

123 Anurag Nagar, Behind Press Complex
Near Jain Temple, INDORE 452011, Madhya Pradesh, India
jain.anilk@gmail.com

We describe, using modern terminology, the encryption and decryption mechanisms used in Ācārya Kumudēndu's remarkable epic *Siribhūvalaya*. We also point out that the verses therein incorporate features for error-detection, similar to the use of parity check bits in modern error-correcting codes. Though the underlying language is primarily Kannaḍa, scholars have pointed out that by varying the patterns used in the decryption method, the same piece of encrypted text can be decrypted to well-known phrases and passages from other languages.

Keywords: *Khaṇḍa, Adhyāya, Cakra, Cakra-Bandha, Navmānk-Bandha*, Substitution-cipher, Transposition-cipher, *Antara Sāhitya*, Steganography.

Mathematics Subject Classification 2020: 01-06, 01A32, 01A99.

*The mathematical formulation of the content and the style of presentation in this paper owe much to the efforts of Prof. Jaikumar Radhakrishnan, Tata Institute of Fundamental Research, Mumbai; but for his reluctance to accept it, the authorship of the paper would have been shared with him. I am grateful to Prof. Radhakrishnan for working on my original draft and bringing it to the present form.

9.1. Introduction

Siribhūvalaya was composed by Ācārya Kumudēndu in the 9th century in the state of Karnataka. It did not receive much attention from *ācāryas* and scholars for several centuries, perhaps owing to the difficulty in deciphering its content. The text covers a vast variety of subjects, and is remarkable in that it is written using a system of numbers, independent of any language specific script, based on a sophisticated symmetric-key encryption and decryption system. Recent efforts, over the past seventy years, have cast some light on the contents of this unique text, leaving readers wonder-struck not only by the multiplicity and profundity of knowledge contained in the text, but also by the novelty and innovation in its presentation and rendition. It is claimed that multilingual poetry and verses, which are derived from decrypted base-text in *Kannaḍa* language, comprise of 718 dialects and languages, including *Prākṛta, Saṁsakṛta, Pāli, Apabhranś, Telugu, Tāmil* and *Marāṭhī.* The subject matter covers many topics of ancient Indian philosophies and sciences, as well as canonical scriptures of Jainism and other prevalent religions at that period. It also includes extensive deliberations on mathematics [1, 2, 5] and *Āyurveda* (specifically *Puṣpāyurveda* and *Lalitāyurveda*).

9.1.1. *Recent scholarship*

A handwritten manuscript of *Siribhūvalaya* on *Korī* paper was in the possession of a Jaina scholar named Dharaṇēndra Paṇdita, who was resident of a village *Doddābalē* near Bengaluru. Yellapā Śāstrī, an Āyurvēdic practitioner and scholar, was able to gain access to this manuscript in 1920. Yellapā Śāstrī spent thirty years with the mysterious of the *Siribhūvalaya*, and was finally able to make a break-through in 1950. In 1953, together with co-editors Karlamaṅgalam Śrīkanṭhaiyā (a freedom fighter with an abiding interest in history and inscriptions, who had been working on the text with Yellapā Śāstrī) and Ananta Subbarāo (inventor of the Kannaḍa typewriter), Yellapā Śāstrī released the first volume of compilations; a second volume was released two years later. In these two volumes, published by the *Sarvārtha Siddhī Saṅgha*, chapters 1 to 33 were decoded and detailed explanations of the contents were provided. In 1956, Yellapā Śāstrī brought some of the transcripts and some original manuscripts to the National Archives of India, New Delhi [1, 5]. These included decoded transcripts of the remaining chapters of the first volume from — chapters 34 to 59. Subsequently, Yellapā Śāstrī worked closely with Ācārya Dēśabhūṣaṇa on a Hindi translation of *Siribhūvalaya.* However, after the sudden demise of Yellapā Śāstrī in 1957, no significant research work or publication seems to have appeared on *Siribhūvalaya* for several decades. Beginning in the first decade of the 21st century, Sudharthī Hassan has written several books in Kannaḍa, compiling the *antarasāhitya* and offering his interpretations. He has thoroughly investigated and compiled the *antarasāhitya* and provided a comprehensive interpretation of the decoded texts of 59 chapters. Working with his own resources, his published works and commentaries on YouTube and other platforms have done much to generate and sustain interest in *Siribhūvalaya* among the Kannaḍa speaking people. In 2012, Narendra Kumar Jain and P.C. Jain published a detailed report, and made valuable contributions in deciphering *Siribhūvalaya* in *Devanāgarī* script, for which they developed special software.

Scholars differ on the period when Ācārya Kumudēndu lived and when *Siribhūvalaya* was composed. Kusum Patoria [4] argues, based on references in the text and external references, that *Siribhūvalaya* was composed in the 9th century; this author considers these arguments very persuasive. Other scholars place the date much later (12th to 15th century) based on linguistic and textual features, but there is insufficient accompanying evidence to support their claims. Ācārya Kumudēndu is believed to have lived in a village at the foot of Nandi Hills near Bengaluru. References to this place have been made by the *Ācārya* himself in his writings, and also in a poem '*Kumudēndu Śataka*' by the poet Devappā. At present this village is known as Yaluvahalli in Chickaballapur district of Karnataka, India.

9.1.2. *Organization of this paper*

In the next section, we describe in detail the structure of the *Siribhūvalaya.* In subsequent sections, we present the methods used in

Siribhūvalaya for decrypting and encrypting information, and touch upon some other features that the text has in common with modern cryptography and error-correction. Finally, we point out directions for further research.

9.2. The Structure of the *Siribhūvalaya*

The *Siribhūvalaya* is organized hierarchically. The text has nine volumes; a volume is referred to as a *Khaṇḍa*. These volumes are named: *ṁagala Prābhṛta, Śrutāvatāra, Sūtravatāra, Prāṇāvaya Pūrva, Dhavala, Jaya Dhavala, Vijaya Dhavala, Mahā Dhavala, Atiśaya Dhavala* [3]. A *Khaṇḍa* in turn consists of several chapters, referred to as an *Adhyāya.* The number of chapters in a volume varies: the first volume, *ṁagala Prābhṛta* has 59 chapters [3]. The number of chapters in the second volume, *Śrutāvatāra*, is not known; only the first chapter is currently available. A chapter is divided further into *Cakras.* The number of *Cakras* in a chapter varies. In the first volume, there are 1263 *Cakras* distributed among its 59 chapters. The *Cakras* carry varying amounts of encrypted information. However, sometimes verses are allowed to straddle *Cakras*, that is, if the last verse does not completely fit into a *Cakra*, the remaining part of that verse is included in the beginning of the next *Cakra.* Current evidence suggests that all *chakras* are 27×27 arrays of numbers in the range $1, \ldots, 64$.

These chapters have been decrypted [6] so far, and they suggest that the information is organized in separate volumes and chapters based on the subject matter. For example, the first volume is general in nature, and includes verses pertaining to various aspects of Jainism, including canonical principles, information on important religious figures, their origins and lives; it refers to other volumes, and contains discussions and interpretation of contemporary philosophies. The fourth volume is devoted to *Puṣpāyurveda*, a system of floral medicine and therapy; it is not publicly available at present. Apart from the first volume, we do not at present have precise information on the number of chapters in other volume. The names of the volumes appear in Yellapā Śāstrī's [6] commentaries. No previous commentary on *Siribhūvalaya*, dating from the several centuries since the composition, is currently available.

In *ṁagala Prābhṛta* only two encryption patterns are used — *Cakra-Bandha* and *Navamānk-Bandha.* In other volumes, other mechanisms are used — e.g., the second volume has a mix of *Bandhas*, some of which involve traversing a 9×9 sub-matrix in a spiral pattern. Names of 40 different *Bandhas* appear in the first volume. In the first volume, a different chapter-key is used for each chapter; it is likely that the same chapter-key is not repeated in other volumes as well.

9.3. Decrypting a *Cakra*

A *Cakra* consists of an 27×27 matrix of numbers in the range $\{1, 2, \ldots, 64\}$. Each of these numbers represents a phonetic unit known as a *Mūla Varṇa.* The general principle followed in the *Siribhūvalaya* is similar to traditional transposition ciphers, where the characters (or some groups of them) are reordered and then presented; to decode such a message, one must, in principle know the inverse of the permutation that was applied to reorder the elements. In *Siribhūvalaya*, the order in which the 729 numbers of the *Cakra* are to be strung together is determined by two keys (i) the **chapter-key**, which is itself represented as a 3×3 matrix of numbers in the range $\{1, 2, \ldots, 9\}$ and (ii) the **sub-matrix-key**, which is represented as a 9×9 or a 27×27 matrix in the published commentaries.[1]

Chapter-Key: A chapter-key is a permutation of the numbers $\{1, 2, \ldots, 9\}$. The chapter key k is written explicitly as a string of nine numbers, $k(1), k(2), \ldots, k(9)$, and arranged in a 3×3 matrix. For example, the following key k^* is used in the second chapter of

[1]Unlike in the case of chapter-keys, we are unable to find the sub-matrix-key explicitly mentioned in the original manuscript; the representation we discuss is borrowed from a relatively recent translation [Ś56].

the first volume.

$$\begin{bmatrix} k^*(1) & k^*(2) & k^*(3) \\ k^*(4) & k^*(5) & k^*(6) \\ k^*(7) & k^*(8) & k^*(9) \end{bmatrix} \equiv \begin{bmatrix} 3 & 4 & 5 \\ 2 & 1 & 6 \\ 9 & 8 & 7 \end{bmatrix}. \tag{1}$$

The decoding using such key proceeds in two stages. First, the *Cakra* matrix *is* partitioned into nine 9×9 sub-matrices as shown below:

$$\begin{bmatrix} S_1 & S_2 & S_3 \\ S_4 & S_5 & S_6 \\ S_7 & S_8 & S_9 \end{bmatrix}$$

where each S_i is a 9×9 matrix. The key $k*$ above specifies that the 9×9 sub-matrices must be decoded in the order S_5, S_4, S_1, S_2, S_3, S_6, S_9, S_8, S_7, and the nine strings, each with 81 numbers, obtained from the individual sub-matrices must be concatenated to obtain the final decoded string. This only partly describes the decoding process; we still need to specify how each of the nine 9×9 sub-matrices is decoded.

Sub-matrix-Key: A common rule, usually specified by a single sub-matrix-key, is used to decode all the sub-matrices within a *Cakra*; in the second volume, there are instances where two sub-matrix-keys are used within a single *Cakra.* Note that a sub-matrix contains 81 numbers and we seek one-to-one correspondence K between the set of indices of a 9×9 matrix, namely, $\{1, 2, \ldots, 9\} X \{1, 2, \ldots, 9\}$ and $\{1, 2, \ldots, 81\}$. To see how this sub-matrix key is used to decode, let K^{-1} denote the inverse of this map (thus, $K^{-1}(L)$ gives us the location in the sub-matrix from where the Lth number in the final decoding should be picked up. Thus, the decoding function with this sub-matrix-key decodes the sub-matrix S as follows:

$$\text{dec}(K, S) = S[K^{-1}(1)]S[K^{-1}(2)] \cdots S[K^{-1}(81)]$$

(Recall that $K^{-1}(1), K^{-1}(2), \ldots, K^{-1}(81)$ are elements of

$\{1, 2, \ldots, 9\} \times \{1, 2, \ldots, 9\}$.)

Decryption: Suppose the chapter-key k and the sub-matrix-key K have been specified. To decode the 27×27 *Cakra* C, one first splits C into its nine sub-matrices $S_1, S_2, \ldots, S_9$ as described above; the final string of numbers is given by

$$\begin{aligned}\mathrm{Dec}((k, K), C) &= \mathrm{dec}(K, S_{k^{-1}(1)}) \circ \mathrm{dec}(K, S_{k^{-1}(2)}) \\ &\quad \circ \cdots \circ \mathrm{dec}(K, S_{k^{-1}(9)}).\end{aligned}$$

The above description covers most of the cases, but there is one exception. Recall that the chapter-key is a 9×9 matrix of numbers, where each of the numbers $1, \ldots, 9$ appears exactly once. Yet, some chapter-keys do not have this structure; in fact, some of their entries are zeroes. We call such keys null-chapter-keys. When the chapter has a null-chapter-key, the decoding proceeds differently. The 27×27 is matrix is decoded together using a single key $K_0 : \{1, 2, \ldots, 729\} \to \{1, 2, \ldots, 27\} \times \{1, 2 \ldots, 27\}$. That is, if the *Cakra* is S, then

$$\mathrm{Dec}((\mathrm{null}, K_0), S) = S[K^{-1}(1)]S[K^{-1}(2)] \cdots S[K^{-1}(729)].$$

We have described above how a *Cakra* is converted to a string of 729 numbers. After this, the individual numbers in this string are replaced by the phonetic units they represent and the final text is obtained by synthesizing them according to Kannaḍa grammar. In this paper, we mainly focus on the textual components of the decryption and encryption processes in *Siribhūvalaya*; however, at the end of this section we briefly comment on some aids for the synthesis step that are already incorporated in the *Cakras*.

9.3.1. *Examples*

In this section, we illustrate the above decryption method using two examples. In the first example, the chapter-key is non-null and the sub-matrix-key corresponds to the 9×9 *Navmānka-Bandha* pattern; in the second, the chapter-key is null, and the sub-matrix-key corresponds to the 27×27 *Cakra-Bandha* pattern.

Example 1. The following example is the first *Cakra* in the second chapter of the first volume. We will need two keys: (i) the chapter-key k^* and (ii) the sub-matrix-key K^*. We have already noted the key for this chapter above, see Eq. (1):

$$(k^*(1), k^*(2), \ldots, k^*(9)) = (3, 4, 5, 2, 1, 6, 9, 8, 7).$$

The sub-matrix key in this case is the following (*Navmānka-Bandha* pattern).

$$K^* \equiv \begin{bmatrix} 47 & 58 & 69 & 80 & 1 & 12 & 23 & 34 & 45 \\ 57 & 68 & 79 & 9 & 11 & 22 & 33 & 44 & 46 \\ 67 & 78 & 8 & 10 & 21 & 32 & 43 & 54 & 56 \\ 77 & 7 & 18 & 20 & 31 & 42 & 53 & 55 & 66 \\ 6 & 17 & 19 & 30 & 41 & 52 & 63 & 65 & 76 \\ 16 & 27 & 29 & 40 & 51 & 62 & 64 & 75 & 5 \\ 26 & 28 & 39 & 50 & 61 & 72 & 74 & 4 & 15 \\ 36 & 38 & 49 & 60 & 71 & 73 & 3 & 14 & 25 \\ 27 & 48 & 59 & 70 & 81 & 2 & 13 & 24 & 35 \end{bmatrix} \tag{2}$$

where the (i, j)th entry is the value $K^*((i, j))$. The numbers have a natural pattern. The traversal starts from the middle cell of the top row, and every successive cell that is visited is located to the 'north-east' of the previous cell. When we fall off the matrix at a boundary, we just wrap-around and continue, as if the matrix were actually written on torus; when a cycle is completed we just move down one cell and continue.

Having specified our keys, we are now ready to look the *Cakra* (see Figure 9.1). To decrypt this *Cakra*, we proceed as stated above and compute

$$\mathrm{dec}(K, S_5) \circ \mathrm{dec}(K, S_4) \circ \mathrm{dec}(K, S_1) \circ \mathrm{dec}(K, S_2) \circ \mathrm{dec}(K, S_3)$$
$$\circ \cdots \circ \mathrm{dec}(K, S_8) \circ \mathrm{dec}(K, S_7)$$

where the S_i refer to the 9×9 sub-matrices as described above. We now describe $\mathrm{dec}(K, S_5)$, the sub-matrix that is decrypted first, in greater detail. The sub-matrix S_5 is the middle 9×9 sub-matrix of

	1	2	3	4	5	6	7	8	9	10	11	12	13	14	15	16	17	18	19	20	21	22	23	24	25	26	27
1	53	1	1	1	60	1	56	1	57	9	59	7	53	1	7	47	35	4	28	25	22	38	13	7	1	1	1
2	33	57	56	40	45	7	52	4	1	1	13	16	35	47	1	1	45	51	59	53	4	45	53	55	47	45	47
3	1	1	2	4	54	1	43	1	1	22	55	7	1	56	44	2	59	54	4	56	1	2	7	7	54	1	1
4	51	52	18	4	53	1	51	56	47	1	47	55	53	1	47	1	1	13	1	58	45	13	48	1	54	59	56
5	1	48	13	1	13	7	16	1	53	1	1	4	52	4	53	16	1	28	7	1	22	1	58	1	4	18	47
6	7	56	55	24	47	30	31	1	56	50	52	47	48	1	45	30	1	7	45	1	1	28	59	54	45	1	54
7	9	1	54	1	4	4	56	1	47	4	1	9	55	4	52	45	47	54	53	45	2	1	1	47	52	1	18
8	7	1	45	54	43	54	54	1	50	47	54	1	56	2	1	1	24	45	7	43	56	45	1	16	60	56	1
9	35	1	28	28	28	2	56	1	47	1	56	1	28	22	54	45	3	4	28	1	47	40	7	7	54	54	54
10	1	16	45	56	1	4	52	1	1	45	56	56	1	2	1	35	56	52	1	4	1	4	28	1	1	47	1
11	56	1	24	13	45	1	56	53	56	1	1	56	57	53	2	1	7	24	54	30	56	47	45	45	16	57	42
12	54	51	22	7	30	1	1	47	55	45	1	4	1	54	43	47	1	1	1	1	1	3	47	47	3	47	1
13	1	45	8	3	52	56	7	3	4	47	43	59	52	4	45	30	52	50	47	56	1	4	1	56	27	45	28
14	1	45	56	1	1	47	59	56	47	1	1	2	46	1	2	1	1	1	1	42	45	44	18	59	43	47	1
15	1	45	59	53	1	1	1	7	46	47	1	2	1	44	55	57	48	1	1	7	1	45	16	4	1	31	43
16	2	1	4	45	13	4	45	47	45	45	59	56	54	1	24	4	53	2	47	52	16	45	54	47	53	4	54
17	54	43	1	1	47	47	1	1	56	53	1	1	54	47	45	4	37	1	59	45	1	1	1	16	55	3	1
18	7	56	54	16	4	50	56	1	59	51	1	4	1	52	45	35	53	1	4	59	53	42	45	1	28	52	4
19	4	7	47	59	50	54	1	52	16	45	55	1	38	13	60	43	53	56	28	1	55	45	53	55	28	1	59
20	56	1	1	1	1	53	16	28	30	1	56	38	1	4	1	1	1	1	43	1	1	9	1	47	13	52	1
21	45	42	56	46	4	38	47	1	1	1	7	30	13	45	55	28	4	45	28	52	51	56	1	18	1	7	4
22	1	1	50	40	58	1	43	54	1	60	1	1	1	1	47	54	47	47	1	1	28	45	45	43	30	42	1
23	1	7	4	16	57	43	48	45	54	30	28	38	56	1	1	4	16	7	56	1	1	47	4	7	50	59	30
24	55	30	13	4	7	1	1	1	45	1	1	9	45	13	30	59	53	47	60	7	1	35	56	52	1	2	7
25	1	1	56	7	56	45	28	7	1	56	51	1	18	3	47	16	1	54	13	50	3	1	16	1	45	54	1
26	1	1	43	1	1	1	60	59	28	42	56	47	51	21	47	51	48	43	4	51	55	24	57	50	4	53	23
27	59	47	42	45	8	1	4	47	50	1	1	4	59	4	7	1	1	1	52	1	56	1	18	16	1	45	54

Figure 9.1. Cakra 1 of volume 1, chapter 2: the shaded region represents the sub-matrix S_5

the Cakra:

$$S_5 \equiv \begin{bmatrix} 45 & 56 & 56 & 1 & 2 & 1 & 35 & 56 & 52 \\ 1 & 1 & 56 & 57 & 53 & 2 & 1 & 7 & 24 \\ 45 & 1 & 4 & 1 & 54 & 43 & 47 & 1 & 1 \\ 47 & 43 & 59 & 52 & 4 & 45 & 30 & 52 & 50 \\ 1 & 1 & 2 & 46 & 1 & 2 & 1 & 1 & 1 \\ 47 & 1 & 2 & 1 & 44 & 55 & 57 & 48 & 1 \\ 45 & 59 & 56 & 54 & 1 & 24 & 4 & 53 & 2 \\ 53 & 1 & 1 & 54 & 47 & 45 & 4 & 37 & 1 \\ 51 & 1 & 4 & 1 & 52 & 45 & 35 & 53 & 1 \end{bmatrix}. \tag{3}$$

The decryption involves reading the numbers in S_5 in the specified by $K*$: the first number in the final sequence is $S_5[(K^*)^{-1}(1)]$, that is, $S_5[1,5] = 2$, the second number is $S_5[(K^*)^{-1}(2)] = S_5[9,6] = 45$, and so on, until $S_5[(K^*)^{-1}(81)] = S_5[9,5] = 52$, to obtain the sequence of 81 numbers:

2 45 4 53 1 1 43 4 57 1 53 1 35 37 2 47 1 59 ... 45 4 48 1 47 1 56 1 52

Finally, we substitute the corresponding phonetic units (shown in Figure 9.2) for each of the numbers in the sequence and obtain the

1	अ	a	17	एा	ē	33	च्	c	49	फ्	ph
2	आ	ā	18	एाा	ē	34	छ्	ch	50	ब्	b
3	आा	ā	19	ऐ	ai	35	ज्	j	51	भ्	bh
4	इ	i	20	ऐी	ai	36	झ्	jh	52	म्	m
5	ई	ī	21	ऐीे	ai	37	ञ्	ñ	53	य्	y
6	ईी	ī	22	ओ	ō	38	ट्	ṭ	54	र्	r
7	उ	u	23	ओो	ō	39	ठ्	ṭh	55	ल्	l
8	ऊ	ū	24	ओोे	ō	40	ड्	ḍ	56	व्	v
9	ऊ्	ū	25	औ	au	41	ढ्	ḍh	57	श्	ś
10	ऋ	r̥	26	औौ	au	42	ण्	ṇ	58	ष्	ṣ
11	ॠ	r̥̄	27	औौै	au	43	त्	t	59	स्	s
12	ॠा	r̥̄	28	क्	k	44	थ्	th	60	ह्	h
13	ऌ	l̥	29	ख्	kh	45	द्	d	61	◌ं	ṁ
14	ॡ	l̥̄	30	ग्	g	46	ध्	dh	62	◌ः	ḥ
15	ॡ्	l̥̄	31	घ्	gh	47	न्	n	63	...	...
16	ए	ē	32	ङ्	ṅ	48	प्	p	64	::	f

Figure 9.2. The table shows the phonetic unit corresponding to each number in the range 1 to 64 in the *Devanāgarī* and the Roman scripts.

following *Kannaḍa* verse:

ādiya atiśaya jñāna sāmrājyada | *sādhita vaybhavavāda* ||

arthāgama avirala śabdava nōōōdipa navam....

Example 2. We now describe how the first *Cakra* of chapter 1 of volume 1 is decrypted (shown in Figure 9.4). The chapter-key for this chapter is null, so this *Cakra* must be decrypted as a 27×27 unit, without breaking it down into 9×9 sub-matrices. This particular *Cakra* has been rendered using the *Cakra-Bandha* pattern, shown in Figure 9.3. The corresponding key, K^C, is similar to the *Navmānk-Bandha* pattern: the middle entry of the top row is 1, that is, $K^C((1, 14)) = 1$, then as one moves up and right one entry, the number increases by 1. For example, from the fact that $K^C((1, 14)) = 1$, it follows that $K^C((0, 15)) = 2$ (we wrap around so $(0, 15)$ is to be thought of as $(27, 15)$) and $K^C((26, 16)) = 3$ and so on. After 27 steps we return to $(1, 14)$, then we move to $(2, 14)$, so that, $K^C((2, 4)) = 27$, and then we continue as before. [In general, every pair (a, b) in the range $1, \ldots, 27\, 1, \ldots, 27$ can be written uniquely in the form $(2i - j, 14 - i + j)$, for (i, j) in $(1, \ldots, 27) \times (1, \ldots, 27)$,

	1	2	3	4	5	6	7	8	9	10	11	12	13	14	15	16	17	18	19	20	21	22	23	24	25	26	27
1	380	409	438	467	496	525	554	583	612	641	670	699	728	1	30	59	88	117	146	175	204	233	262	291	320	349	378
2	408	437	466	495	524	553	582	611	640	669	698	727	27	29	58	87	116	145	174	203	232	261	290	319	348	377	379
3	436	465	494	523	552	581	610	639	668	697	726	26	28	57	86	115	144	173	202	231	260	289	318	347	376	405	407
4	464	493	522	551	580	609	638	667	696	725	25	54	56	85	114	143	172	201	230	259	288	317	346	375	404	406	435
5	492	521	550	579	608	637	666	695	724	24	53	55	84	113	142	171	200	229	258	287	316	345	374	403	432	434	463
6	520	549	578	607	636	665	694	723	23	52	81	83	112	141	170	199	228	257	286	315	344	373	402	431	433	462	491
7	548	577	606	635	664	693	722	22	51	80	82	111	140	169	198	227	256	285	314	343	372	401	430	459	461	490	519
8	576	605	634	663	692	721	21	50	79	108	110	139	168	197	226	255	284	313	342	371	400	429	458	460	489	518	547
9	604	633	662	691	720	20	49	78	107	109	138	167	196	225	254	283	312	341	370	399	428	457	486	488	517	546	575
10	632	661	690	719	19	48	77	106	135	137	166	195	224	253	282	311	340	369	398	427	456	485	487	516	545	574	603
11	660	689	718	18	47	76	105	134	136	165	194	223	252	281	310	339	368	397	426	455	484	513	515	544	573	602	631
12	688	717	17	46	75	104	133	162	164	193	222	251	280	309	338	367	396	425	454	483	512	514	543	572	601	630	659
13	716	16	45	74	103	132	161	163	192	221	250	279	308	337	366	395	424	453	482	511	540	542	571	600	629	658	687
14	15	44	73	102	131	160	189	191	220	249	278	307	336	365	394	423	452	481	510	539	541	570	599	628	657	686	715
15	43	72	101	130	159	188	190	219	248	277	306	335	364	393	422	451	480	509	538	567	569	598	627	656	685	714	14
16	71	100	129	158	187	216	218	247	276	305	334	363	392	421	450	479	508	537	566	568	597	626	655	684	713	13	42
17	99	128	157	186	215	217	246	275	304	333	362	391	420	449	478	507	536	565	594	596	625	654	683	712	12	41	70
18	127	156	185	214	243	245	274	303	332	361	390	419	448	477	506	535	564	593	595	624	653	682	711	11	40	69	98
19	155	184	213	242	244	273	302	331	360	389	418	447	476	505	534	563	592	621	623	652	681	710	10	39	68	97	126
20	183	212	241	270	272	301	330	359	388	417	446	475	504	533	562	591	620	622	651	680	709	9	38	67	96	125	154
21	211	240	269	271	300	329	358	387	416	445	474	503	532	561	590	619	648	650	679	708	8	37	66	95	124	153	182
22	239	268	297	299	328	357	386	415	444	473	502	531	560	589	618	647	649	678	707	7	36	65	94	123	152	181	210
23	267	296	298	327	356	385	414	443	472	501	530	559	588	617	646	675	677	706	6	35	64	93	122	151	180	209	238
24	295	324	326	355	384	413	442	471	500	529	558	587	616	645	674	676	705	5	34	63	92	121	150	179	208	237	266
25	323	325	354	383	412	441	470	499	528	557	586	615	644	673	702	704	4	33	62	91	120	149	178	207	236	265	294
26	351	353	382	411	440	469	498	527	556	585	614	643	672	701	703	3	32	61	90	119	148	177	206	235	264	293	322
27	352	381	410	439	468	497	526	555	584	613	642	671	700	729	2	31	60	89	118	147	176	205	234	263	292	321	350

Figure 9.3. The *Cakra-Bandha* key K^C.

where we perform arithmetic modulo 27 with numbers in the range $(1, \ldots, 27)$. Then, $K^C((a,b)) = 27.\ (I-1) + j$.]

When the *Cakra* of Figure 9.4 is decrypted using the key K^C, we obtain the following sequence of 729 numbers, each corresponding to a phonetic unit: 1 58 38 1 52 ... 60 7 41 56 7. The first 96 of these units when synthesized give us:

aṣṭa mahāprātihārya vaybhavadinda | *aṣṭaguṇangaḷōḷ ōmdam* ||
srṣṭigē mangaḷa paryāyadinita | *aṣṭama jinagēra guvēnu* ||1||

	1	2	3	4	5	6	7	8	9	10	11	12	13	14	15	16	17	18	19	20	21	22	23	24	25	26	27
1	59	23	1	16	1	28	28	1	1	56	59	4	56	1	1	47	16	34	1	7	16	1	1	7	56	1	60
2	53	54	47	28	1	47	45	28	7	4	59	41	4	45	1	30	47	47	45	42	53	28	51	1	52	1	1
3	1	22	1/5	30	2	1	2	55	30	1/7	7	45	47	52	1	4	1	47	1	1	1	1	53	1	52	59	52
4	59	30	2	55	55	13	16	2	53	60	1	4	16	47	48	45	16	56	56	43	45	1	56	1	4	1	13
5	47	45	1	1	22	30	51	1	2	56	38	30	4	1	1	56	1	1	16	1	57	7	56	56	1	22	1
6	54	52	52	45	1	7	55	48	1	58	52	35	28	55	1	38	45	30	55	4	47	7	45	38	45	38	1
7	1	1	1	28	13	56	55	51	54	1	1	1	1	42	2/2	4	4	1	43	16	47	7	1	13	4	51	4
8	28	53	47	22	8	1	53	59	38	7	43	40	1	52	59	54	30	1	45	16	1	28	23	50	7	43	43
9	1	2	45	51	30	1	52	58	48	59	47	54	4	4	1	47	45	47	56	28	1	45	1	13	7	7	7
10	55	1	53	47	56	1	1	7	1	1	2	60	48	56	1	1	16	1	1	54	1	52	17	30	54	45	45
11	59	56	52	1	45	1	55	28	52	28	1	2	1	52	54	4	43	60	48	28	1	16	23	8	53	7	1
12	2	1	53	52	43	23	2	4	16	52	44	54	1	2	42	7	1/4	7	47	30	28	48	47	1	54	52	16
13	45	54	23	4	28	45	45	30	1	59	1	56	28	2	54	53	38	2	2	1	28	55	40	60	4	50	28
14	2	13	47	1	1	4	17	45	1	56	1	52	56	51	1	47	55	55	45	7	2	54	1	56	7	1	1
15	23	4	53	54	59	48	13	56	1	47	23	1	2	55	16	1	1	47	40	54	16	52	1	47	60	43	60
16	45	16	43	1	7	47	1	7	1	4	54	54	1	43	28	28	7	1	2	7	52	30	1	4	47	4	13
17	42	1	54	13	1	28	1	45	42	5	48	56	1	1	1	52	54	7	1	1/6	2	56	56	2	43	1	1
18	56	43	22	45	56	43	2	2	56	1	8	48	59	59	7	16	53	55	53	48	1	1	46	2	30	53	1
19	47	45	1	2	54	56	56	2	55	51	4	16	7	13	30	16	1	1	4	52	52	4	54	47	2	38	1
20	1	54	60	56	54	1	60	1	1	16	40	38	17	1	47	56	33	55	1	1	59	48	1	53	7/1	1	1
21	1	52	16	1	60	1	30	53	30	7	47	13	13	22	8	13	45	59	54	1	2	42	54	47	53	52	53
22	16	30	1/3	4	52	47	56	1	28	16	1	22	59	51	1	1	7	28	53	60	7	1	16	16	1	1	58
23	4	53	56	1	52	2	13	52	38	30	45	7	1	30	56	16	1	1	1	30	48	56	54	54	55	28	45
24	1	47	47	1	28	22	1	47	1	1	45	46	1	1	47	53	55	52	1	1	7	43	2	1	1	1	43
25	1	4	53	1	45	43	16	55	52	4	47	55	45	22	51	56	1	38	13	30	2	28	56	13	56	28	55
26	4	16	46	1	1	16	1	1	1	1	1	47	59	4	8	38	58	1	1	48	1	7	22	1	1	1	60
27	52	4	30	56	53	52	54	1	30	52	1	16	54	7	58	1	30	54	1	56	51	53	56	57	56	4	60

Figure 9.4. *Cakra* 1 of volume-1, chapter 1: when two integers appear in the same cell, e.g. in cell $(3, 3)$, we have 1/5 to indicate that both 1 and 5 appear in that cell.

In this example (see Figure 9.4), indications where a verse ends are provided by markers in the form of composite symbols. For example, the 96 phonetic units comprising the verse reproduced above extend from position (1, 14) up to position (20, 25). In position (20, 25), we find 7/1, where 7 corresponds to the last phonetic unit of the verse, and 1 indicates that the first verse has just ended. Note that the last verse ends at position (3, 10); the subsequent 32 numbers when decoded yield meaningful words, but they are not included in the decryption provided by Yellapā Śāstrī [6]. Furthermore, the verses themselves often have different lengths and meters: the shorter *Pāda-pada*, which is about one-fourth the length of the longer *Pūrṇa-pada*.

9.4. Encryption (Encoding) Method

The encryption method naturally complements the decryption (decoding) method, it essentially proceeds in reverse.

The verse that is sought to be encrypted is first broken into phonetic units; each phonetic unit is encoded using a number in the range $\{1, 2, \ldots, 64\}$. Assume that this results in a string M of 729 integers. As before we have a chapter-key k ; let us assume that it is not null. To obtain the encryption of M, a 27×27 matrix, we proceed as follows:

Step 1: Let $M = M_1 M_2 \ldots M_9$, where each M_i is string of 81 numbers.

Step 2: Using the sub-matrix key Kobtain a sub-matrix $S_i = \mathrm{enc}(K, M_i)$ from M_i as follows: $S_i[a, b] = M_i[K((a, b))]$, for $a, b \in \{1, \ldots, 27\}$.

Step 3: Arrange the 9 matrices S_i in a 3×3 grid using the order specified by k. Thus the encryption of M is given by the following 27×27 matrix:

$$\mathrm{Enc}((\kappa, K), M) = \begin{bmatrix} S_{\kappa^{-1}(1)} & S_{\kappa^{-1}(2)} & S_{\kappa^{-1}(3)} \\ S_{\kappa^{-1}(4)} & S_{\kappa^{-1}(5)} & S_{\kappa^{-1}(6)} \\ S_{\kappa^{-1}(7)} & S_{\kappa^{-1}(8)} & S_{\kappa^{-1}(9)} \end{bmatrix}.$$

When the chapter key k is null, we proceed similarly, but do not carry out Step 3. Note that in either case, we have $\mathrm{Dec}((k, K), \mathrm{Enc}((k, K), M)) = M$.

9.5. Error Detection at the End of Each Chapter

The last verse of every chapter contains the total number of phonetic units in the chapter, from which we can derive the number the chakras in the chapter. In addition, the total number of derived texts (which can be obtained from the *Cakra*s in the chapter by using alternative patterns of reading) are also mentioned in the last verse. Yellapā Śāstrī noticed and documented this [5, 6].

9.6. Multiple Languages

Antarasāhitya is mentioned in the original text at many places [2, 3, 5]. In the Jaina tradition, discourses/scriptures are expected to be understood in multiple languages. References to that tradition exist in *Siribhūvalaya*. Research articles have discovered other examples. Some believe that lost works in the Jaina tradition may be recovered from the *Siribhūvalaya*.

9.7. Related Concepts in Modern Cryptography

Encryption and decryption using keys (shared by the sender and receiver) is fundamental to modern cryptography. Main cryptographic mechanisms used in *Siribhūvalaya* are substitution-cipher (e.g., replacing phonetic units by numbers) and transposition-cipher (i.e., reordering of the sequence of numbers based on a key). A common substitution method for phonetic units is used, so it is unlikely to have provided much security. Using transpositions for cryptographic purposes has a long history: ciphers have been used for over two thousand years, and clever devices were used for encryption and decryption. In modern cryptographic protocols the security depends on the key being picked randomly from a large set. In *Siribhūvalaya*, keys vary from chapter to chapter. At least 40 different sub-matrix keys are used in *Siribhūvalaya*. In the first volume, a different chapter-key is for different chapters; it is likely that chapter-keys are not repeated in other volumes as well. By modern standards the set of keys is probably very small, and it is unlikely to have provided robust security; however, the underlying principle seems to have been similar. It would be useful to analyze the frequency with which various chapter-keys and sub-matrix keys are used over the entire text in order to determine if they were, in fact, chosen randomly.

There are some other features of *Siribhūvalaya* that parallel more recent developments in modern computer-based cryptography. In some chapters, the *Cakras* also include *antarasāhitya* or secondary text, apart from the primary text. To obtain these texts, one picks syllables (e.g., the first syllable, middle syllable, sometimes in reverse order) of the various verses and strings them together. In many cases, this results in a well-known and meaningful text in another language, typically, the commonly used *Prākṛta* language (the most popular medium for the transmission of knowledge in the Jaina tradition). In Jaina tradition, discourses/scriptures are expected to be understood in multiple languages. Research articles have discovered several examples to show that this practice has influenced the composition of *Siribhūvalaya* itself. Thus the same cipher text yields multiple alternative texts in different languages based on how it is decrypted; this is loosely related to recent notions of attribute-based cryptography and functional encryption. Interweaving of antarasāhitya in encoded form is related to the modern cryptographic concept of Steganography. However, we do not at present know the adversary from whom the information in the text was sought to be kept secret or some other motivation or need for the author of *Siribhūvalaya* to encrypt information. The scope for research on subjects presented by *Siribhūvalaya* is vast; we hope that this article will spur further interest.

Acknowledgments

The author is thankful to the referees — Dr. Shrikrishna G. Dani and Dr. Jaikumar Radhakrishnan for their valuable comments and suggestions. Author expresses deep gratitude toward Dr. Surender K. Jain for his recommendations and initiatives in involving Dr. Jaikumar Radhakrishnan in compilation of this paper. Author is also thankful to organizers of "Symposium on Jainism and

Mathematics" for providing a platform to present author's finding on *Siribhūvalaya.*

References

[1] Anil Kumar Jain, The Siribhūvalaya: An unexplored treasure trove of knowledge and creativity, Newsletter of the Centre of Jaina Studies, X(15), 45–48, 2017.
[2] Anil Kumar Jain, Apratim Pratibhā Sindhu Siri Bhūvalaya Rachyitā Ācārya Kumudēndu (in Hindī), Siri Bhūvalaya Vak Pīṭha, Indore, 2021.
[3] Anil Kumar Jain and Anupam Jain, Siri Bhūvalaya Śodha: Daśā evaṁ Diśā, Kundakunda Jñānapīṭha, Indore, 2017.
[4] Kusum Patoria, Svayambhū Kē Pa'umacari'u Kē Dō Viśiṣta Ullēkha. In Apabhranś Bhāratī, pages 1–4, Apabhranś Sāhitya Akādamī, October 2015–2016.
[5] Swarna Jyothi, Siri Bhūvalaya (Hindi, Vol. 1), Pustak Shakti Publication, Bangalore, 2008.
[6] Yellapā Śāstrī, Bhūvalaya MSS (in Kannaḍa), Vol. 1. National Archives of India, New Delhi, 1956.

https://doi.org/10.1142/9789811255502_0010

Article 10

Jain Syllogism and Other Aspects of Jain Logic with Applications to Statistical Thinking

Kanti V. Mardia

Department of Statistics, University of Leeds
Leeds, LS2 9JT, UK
k.v.mardia@leeds.ac.uk
Department of Statistics, University of Oxford
Oxford, OX1 3LB, UK
mardia@stats.ox.ac.uk

Anthony J. Ruda

Jain Noble Truths Association
2021–2022 Bhagwan Mahavira Fellow
International School for Jain Studies,
844 BMCC Road Shivajinagar,
Pune – 411 004, Maharashtra, India
ajr2192@columbia.edu

It is well known that the core concept of Jain logic, the conditional holistic principle (known as *Anekāntavāda*), originated in the ancient Jain literature. From this core concept, the conditional predication principle (known as *Syādvāda*) was developed and has since become one of the important areas of ancient Jain logic in which some aspects have been well studied in modern terminology.

However, some aspects of Jain logic have not yet been explored and are dealt with in this article. This article gives historical details on the five-fold Jain syllogism (known as *Pañcāvayavavākya*) as part of a comprehensive analysis, thereby unifying the literature and bringing the concept in line with modern mathematical terminology. Its five components — proposition (*Pratijñā*), reason (*Hetu*), example (*Udāharaṇam*), application (*Upanaya*), and conclusion (*Nigamanam*) — are discussed in depth to show how *Pañcāvayavavākya* matches the basis of the current statistical inference. In particular, *Syādvāda* in combination with *Pañcāvayavavākya* yields the Bayesian approach.

Further, a formal and succinct notation is introduced to describe *Syādvāda* and to illustrate the modern mathematical connections, while *Anekāntavāda* is shown to have as its corollary a stratified sampling.

Finally, an element of *Syādvāda* inherent to Turing's cryptographic work towards breaking the Enigma code — work that could fairly be called 'enigmatic statistics' — is revealed.

Keywords: ancient Jain logic, Aristotle, Bayesian analysis, conditional holistic principle (*Anekāntavāda*), conditional predication principle (*Syādvāda),* five-fold Jain syllogism (*Pañcāvayavavākya*), sampling, Turing.

Mathematics Subject Classification 2020: 01-06, 01A32, 01A99, 03A05, 62A01

10.1. Introduction

We analyse some overlooked aspects of ancient Jain logic primarily, and draw applications to present-day statistical thinking secondarily. In particular, we go beyond the well-known three-fold syllogism of Aristotle to give an historical account of the five-fold Jain syllogism (the central topic of this paper) and show that Bhadrabāhu played a major role in its development. We then make this history dynamic by presenting statistical applications to add modern relevance, thereby uniting past and present.

While some aspects of ancient Jain logic have already been interpreted in relation to statistics [6, 12], quantum theory [11], and Boolean algebra [9, 23], other aspects remain unexamined in the modern mathematical context. Broadly, our analysis of various aspects of ancient Jain logic (e.g., *Anekāntavāda*, *Syādvāda* and *Pañcāvayavavākya*) reveals how they are qualitatively figured into Bayesian analysis, stratified sampling, and Alan Turing's 'enigmatic statistics'.

A summary of ancient Jain logic as given in the original scriptures can be found in *A History of Indian Logic* by S.C. Vidyabhusana [29]. More recent texts such as *The Central Philosophy of Jainism (Anekānta-vāda)* by B.K. Matilal [19] and *Jain Philosophy: Historical Outline* by N.N. Bhattacharyya [1] also explore ancient Jain logic in the context of Jain philosophy in general. These texts and others have been referenced in giving the historical account and may be of general interest beyond the scope of this paper. Similarly, for a grounding in the statistical aspects of this paper, a very good introductory text on Bayesian analysis is *Bayesian Data Analysis* by A. Bolstad, *et al.* [2], and a review of stratified sampling is included in *Sampling* by S.K. Thompson [27]. For Turing's statistical work, see his wartime papers such as "The Applications of Probability to Cryptography" [28], now available via *The National Archives* after being declassified in 2012.

10.2. Materials and Methods

We examine the cornerstones of ancient Jain logic — *Anekāntavāda*, *Syādvāda* and *Pañcāvayavavākya* — in the context of modern statistics by advancing a formal notation for the *Syādvāda* system of conditional predication and utilising a Venn diagram [4] to illustrate the modern connections.

10.3. Results and Discussion

10.3.1. *Five-fold Jain syllogism (पञ्चावयववाक्य/Pañcāvayavavākya)*

We begin with the five-fold Jain syllogism. Recall that a syllogism is a form of logical argument consisting of a series of propositions from which a conclusion may be deduced. It should be noted that, as is the case with translating *Pañcāvayavavākya* as the five-fold Jain syllogism, we have given first- or second-order approximate translations for many of the terms used in this paper. *Pañcāvayavavākya* itself is a compound word in Sanskrit, and we can see its literal meaning by breaking it down as follows:

पञ्च/*pañca* = five

+

अवयव/*avayava* = limb or component

+

वाक्य/*vākya* = word or speech.

A first-order translation might therefore be "five-limbed speech." The term for a general syllogism, *Nyāyavākya*, similarly is a compound word in Sanskrit:

न्याय/*nyāya* = method or rules

+

वाक्य/*vākya* = word or speech (as above).

Incidentally, Mahāprajña [13] instead uses the term *mahāvākya* for a general syllogism. In any case, the five-fold Jain syllogism is detailed in the *Tarka-Sangraha* of Annambhatta [26] as consisting of five components:

प्रतिज्ञा/*Pratijñā*, हेतु/*Hetu*, उदाहरणम्/*Udāharaṇam*, उपनय/*Upanaya*, निगमनम्/*Nigamanam*

प्रतिज्ञाहेतूदाहरणोपनयनिगमनानि पञ्चावयवाः ।।

Pratijñāhetūdāharaṇopanayanigamanāni Pañcāvayavāḥ.

A specific example from this scripture [26] relates to fire on a mountain, which we give in the original Sanskrit, its transliteration, and a translation into English:

पर्वतो वह्निमानिति प्रतिज्ञा । धूमवत्त्वादिति हेतुः । यो यो धूमवान् स स वह्निमान् यथा महानसमित्युदाहरणम् । तथा चायमित्युपनयः । तस्मात्तथेति निगमनम्।।

Parvato Vahnimāniti Pratijñā... Dhūmavattvāditi hetuḥ... Yo Yo Dhūmavān Sa Sa Vahnimān Yathā Mahānasamityudāharaṇam... Tathā cāyamityupanayaḥ... Tasmāttatheti nigamanam

(1) There is a fire on the mountain
(2) because there is smoke on the mountain;
(3) where there is smoke, there is fire, as in the kitchen;
(4) since there is smoke on the mountain,
(5) there is a fire on the mountain.

We analyse this example in terms of the five components *Pratijñā*, *Hetu*, *Udāharaṇam*, *Upanaya*, and *Nigamanam* as follows:

(1) *Parvato Vahnimāniti Pratijñā/There is a fire on the mountain —*
Pratijñā — the proposition/hypothesis to be proved.

(2) *Dhūmavattvāditi hetuḥ/Because there is smoke on the mountain —*
Hetu — the reason, cause, or evidence for the hypothesis, as the indication of the basis of the conclusion.

(3) *Yo Yo Dhūmavān Sa Sa Vahnimān Yathā Mahānasamityudāharaṇam/Where there is smoke there must be fire, as seen in the kitchen —*
Udāharaṇam — the example establishing the general rule: the kitchen is symbolic of all places where there is fire.

(4) *Tathā cāyamityupanayaḥ/The mountain is smoky —*
Upanaya — the application of the rule, as the basis of the argument.

(5) *Tasmāttatheti nigamanam/And therefore there is a fire on the mountain —*
Nigamanam — the conclusion, as fire and smoke are both observed to exist as cause and effect.

Accordingly, we will use the terminology below in translating the five components:

Pratijñā =proposition,
Hetu =reason,
Udāharaṇam = example,
Upanaya = application,
Nigamanam =conclusion.

We see that, with the exception of (3), the components of the Jain syllogism can be paired as (1,5) and (2,4):

(1), the postulation of the proposition to be proved,

pairs with

(5), the statement that (1) has been proved,

and

(2), the indication of the basis of the conclusion

pairs with

(4), the application of the basis (2) in the argument itself.

This was further analysed by J. M. Rogers and M. K. Jain [25]:

> "This inference strategy permits a transition between the general principle that has been extracted from empirical data to the specific use (extrapolation) of the principle for the current experience. The 'principle' invoked in this traditional example is deliberately weak. This is also the strength of the Nyāya schema as it forces consciousness of the fact that the conclusion is based on a particular example: 'Wherever there is smoke, there is fire, as in the kitchen.' Although the procedure may strengthen the initial hypothesis, one is not allowed to forget the limits of the knowledge base and the liabilities of the conclusion. It is through iterations of the stimulus-inference-verification cycle (by using different examples) that the degree of certainty is increased. Knowledge, then, is a formalism of past experience and derives its authority from nothing else!"

Indeed, the syllogism is said to be accurate when all five of its components are in harmony with each other, and otherwise constitutes a fallacy (*ābhāsa*) if any of these five parts are discordant with our observations (see Mardia [16]). As Mookerjee [20] notes, "Logic has to work upon the data of experience and is as much an instrument as experience is... Logic is to rationalize and systematize what experience offers."

In another example, the five components are:

(1) William will die (proposition);
(2) because he is a man (reason);
(3) like Jack, Fox, and Herbert (examples);
(4) as they died (application);
(5) therefore, William will die (conclusion).

This example was rearranged by J.L. Jaini [10] as follows:

(1) Jack died, Fox died, and so did Herbert.
(2) Jack, Fox, and Herbert are truly universal types of men.
(3) Therefore, all men die.
(4) William is a man.
(5) Therefore, William will die.

While a 'medium' syllogism in Jain logic consists of five components (see Mardia [16]), we see that the last three terms of the medium syllogism are easily recognised as the three components (major premise, minor premise, and conclusion) of the Aristotelian syllogism:

(1) All men are mortal (major premise).
(2) William is a man (minor premise).
(3) Therefore, William is mortal (conclusion).

This, of course, can be abstracted to a more general form, where M=Middle, S=Subject, and P=Predicate terms of the argument:

Major premise: All mortals die (All M are P).
Minor premise: All men are mortal (All S are M).
Conclusion: All men die (All S are P).

10.3.2. *History and development of the five-fold Jain syllogism*

In the previous section, we have shown the connection between the five-fold Jain syllogism and the three-fold Aristotelian syllogism. Indeed, Vidyabhusana [29] explored whether the five-fold syllogism and the Aristotelian syllogism may have a common origin. Leaving aside the difficulties of establishing such an origin, what is known is that the components of the five-fold syllogism are given in Akṣapāda's *Nyāya Sūtra* [29], and mention of such a syllogism can even be found in the *Mahābhārata* in a passage referring to "a speech of five parts" [29] (recall the literal meaning of *Pañcāvayavavākya* given in Section 10.3.1). This would suggest a Vedic origin, but as Vidyabhusana [29] remarks, it also establishes that while Akṣapāda may have introduced the five-fold syllogism in his *Nyāya Sūtra*, "he was by no means the first promulgator of the doctrine — nay, not even its first disseminator."

On the other hand, Bhadrabāhu advanced a ten-fold syllogism (*daśāvayava-vākya*) in his *Dasavaikālikaniryukti* [29]. The ten components are *pratijñā* (proposition), *pratijñā-vibhakti* (limitation of the proposition), *hetu* (reason), *hetu-vibhakti* (limitation of the reason), *vipakṣa* (counter-proposition), *vipakṣa-pratiṣedha* (opposition to the counter-proposition), *dṛṣṭānta* (example), *āsaṅkā* (validity of the example), *āsaṅkā-pratiṣedha* (meeting of the question), and *nigamana* (conclusion) [1, 29]. For a time, this ten-fold syllogism was held as the standard before the five-fold syllogism gained wide acceptance, at least among Jain logicians: Bhattacharyya [1] notes that "the writings of Candraprabha Sūri and Ratnaprabha Sūri... characterised the syllogism of ten parts as the best (*uttama*), of five parts as the mediocre (*madhyama*) and of two parts as the worst (*jaghanya*)." Given this history, it is an open question as to whether the five-fold syllogism was derived from an earlier ten-fold syllogism, as Bhattacharyya suggests, or if Bhadrabāhu instead took the five-fold syllogism as a basis and extended it to ten components. Nevertheless, the ten-fold syllogism is held as a uniquely Jain contribution to logic [1, 29].

It must be emphasised that Vedic, Jain, and Buddhist philosophies did not develop separately in a vacuum, and so establishing the uniqueness of any single facet of Indian logic can be very difficult. Regardless of scholarly debate on Jainism's pre-Vedic origins, Mahāvīra nevertheless brought pre-existing Jainism to greater prominence, and as he was contemporaneous with the Buddha, subsequently there was much commentary and debate between these three traditions [19]. Vidyabhusana remarks [29]:

> "Since the Brāhmaṇas did not in respect of their social practices differ so markedly from the Jainas as they did from the Buddhists, their attack on the Jaina Logic was not so violent as that on the Buddhist Logic. In fact the logical theories of the Jainas are in many instances akin to those of the Brāhmaṇas. The [technical terms] of the Sthānāṅga-sūtra and the Sūtra-kṛtāṅga of the Jainas are in their meanings similar to, if not altogether identical with, the corresponding terms of the Caraka saṃhitā and the Nyāya-sūtra of the Brāhmaṇas. There was in the Logic of the Brāhmaṇas a casual review of the syllogism of ten members as propounded in certain works of the Jainas, but there was no protracted quarrel on that account between the two parties. The Jaina logicians quoted Brāhmaṇic authors generally in an academic spirit. The special Jaina doctrines of *Naya* (method) and *sapta-bhaṅgī* (sevenfold paralogism), though occasionally criticised, did not receive any rude blows from the Brāhmaṇas."

Beyond the origins of the five-fold syllogism itself, it is also interesting to note the development of its second component, *hetu*. The term can be found in Jain scriptures dating back to the *Bhagavati Sūtra* and *Sthānāṅgasūtra* [29], but had not yet acquired a distinct meaning: as Vidyabhusana [29] notes, "In the Sthānāṅgasūtra, it is used not only in the sense of reason, but also as a synonym for valid knowledge (*Pramāna*) and inference (*Anumāna*)."

Bhattacharyya [1] notes that *hetu* was perhaps first used exclusively in the sense of reason in Bhadrabāhu's ten-fold syllogism as detailed in the *Dasavaikālikaniryukti*, but was ascribed greater significance only by later logicians (such as Hemacandra, as we will soon see), as the middle term within the five-fold syllogism. Looking again at the earlier example, we see that "smoke" is the middle term, as it cannot occur outside of its connection with the major term ("fire"), and is also connected to the minor term ("mountain"):

(1) There is a fire on the mountain;
(2) because there is smoke on the mountain;
(3) and where there is smoke, there is fire.

In other words, the mountain (minor term) is on fire (major term) because it is full of smoke (middle term).

Hetu, which we have earlier defined as the reason given for a hypothesis, has also been translated as "probans" (derived from the Latin verb meaning "to demonstrate" or "to prove"), as in Tatia's translation of Acharya Mahāprajña's *Jaina Nyāya Kā Vikāsa* [13] and Mookerjee and Tatia's translation of Hemacandra's *Pramāṇa-mīmāṃsā* [21]. Tatia [13] translates our "five-fold" syllogism as "five-limbed" — those limbs being the thesis, probans, example, application, and conclusion, respectively — while Hemacandra [21] defined the syllogism and each of its terms in aphorisms XI through XV of the *Pramāṇa-mīmāṃsā* as follows:

XI. The thesis is the statement of the theme to be proved.
XII. Statement of a probans ending in inflexion unfolding the character of probans is (called) the reason.
XIII. Illustration is the statement of the example.

XIV. Application is the act of bringing the probans into connection with the subject.

XV. Conclusion is (the predication of) the probandum.

10.3.3. *Five-fold syllogism in Bayesian terminology*

Let us revisit the five-fold Jain syllogism example given at the end of Section 10.3.1, which can be viewed in Bayesian terminology by combining with *Syāt* from the conditional predication principle *Syādvāda.* We will translate *Syāt* loosely as "maybe" for the time being, while a fuller exposition on *Syādvāda* itself will follow in Section 10.3.4. In Bayesian terms, we have the following:

(1) *Pratijñā*: Jack died, Fox died, and so did Herbert (data).
(2) *Syāt-Hetu*: "Syāt" Jack, Fox and Herbert are truly universal types of men (prior probability).
(3) *Syāt-Udāharaṇam*: "Syāt" all men die (posterior probability).
(4) *Upanaya*: William is a man (new observation).
(5) *Syāt-Nigamanam*: "Syāt" William will die (prediction probability).

This is illustrated qualitatively in the flow chart in Figure 10.1 below.

Specifically in terms of Bayes' Theorem, however, we have event **B** as the data; namely, the deaths of Jack, Fox, and Herbert; and **A** as the event of William's death given the observation that he is alive:

$$P(A|B) = \frac{P(B|A)P(A)}{P(B)}$$

P(**A**|**B**) is the posterior probability, or the probability of **A** being true given that **B** is true.

P(**B**|**A**) is the likelihood, or the probability of **B** being true, given **A** is true.

P(**A**) is the prior probability, or the probability of **A** being true.

P(**B**) is the marginal probability of the data (overall probability unconditional of any event).

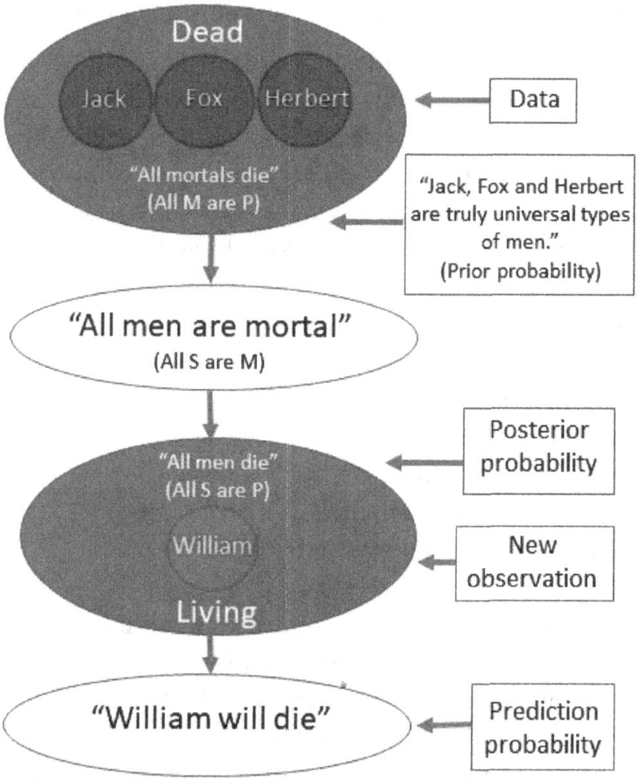

Figure 10.1. Flow chart of the five-fold Jain syllogism

The above example of a medium syllogism has a close similarity to the general form (1) All M are P, (2) All S are M, (3) All S are P, where M=Middle, S=Subject, and P=Predicate, and can also be analysed in Bayesian terminology.

It is clear that the medium syllogism combines inductive and deductive methods of reasoning. In fact, this reflects the main stages of scientific/statistical thinking: the first two terms can be thought of as taking observations from a population and the third term as drawing inference from the observations, while the last two terms give a projection about a new observation. This empirical logic is the basis of scientific methods and should not be lost sight of in all statistical applications (see Mardia [16]).

10.3.4. *Conditional predication principle (Syādvāda)*

In our examination of the five-fold Jain syllogism, we introduced another central feature of Jain logic: the principle of conditional predication (*Syādvāda*; a first-order approximate translation may be "the assertion of possibilities" [29]), by which one examines inference from seven standpoints (*Saptabhaṅgī-naya*), each prefixed by Syāt, which we had previously translated loosely as "maybe":

(1) it is (from one standpoint);
(2) it is not;
(3) it is and is not;
(4) it is indeterminate;
(5) it is and is indeterminate;
(6) it is not and is indeterminate;
(7) it is, is not, and is indeterminate.

This uniquely Jain system of logic was first articulated in Bhadrabāhu's *Sutrakṛtanganiryukti* and treated comprehensively in Samantabhadra's *Āptamīmāṁsā* [29]. It should be noted that "maybe" is not a very good translation of the word "syāt", as this would imply a possible resolution into "maybe so" or "maybe not", and contributes to a sense of confusion as to whether *Syādvāda* entails a kind of probability, as argued by Mahalanobis [12], or something more deterministic — an issue raised by L.C. Jain [7], who, alongside C.K. Jain [8], argued that "indeterminacy and uncertainty are two different aspects" and that the early Jain logicians were actually more inclined towards a "philosophical attitude [of] determinacy". As such, a more accurate translation than "maybe" is "from one standpoint", which we have introduced above.

As another point of clarification, regarding Predication 4 ("it is indeterminate"), Mahalanobis [12] has commented:

> "The fourth category is *avaktavya* which I have translated as 'indeterminate'. Other authors have used the words 'indescribable', or 'inexpressible' or 'indefinite'. For example, Satkari Mookerjee [20] explains 'The inexpressible may be called indefinite'... I prefer

'indeterminate' because this is nearer the interpretation I have in mind."

We have followed with Mahalanobis' translation.

Indeed, this topic has been studied extensively in the last century, by statisticians (Mahalanobis [12], Haldane [6]) and logicians (Matilal [19]) as well as in the context of Boolean algebra (Ramachandran [23], M.K. Jain [9]) and quantum physics (Kothari [11]), to give a few examples.

In a broad mathematical context, a succinct notation can summarise the concept quite well. A very ancient and good example of conceptual notation is for zero, which originated from the Sanskrit *śūnya*, while a modern example is the derivative notation, d/dx. There has not been any difficulty in assigning "+" to Predication 1 ("it is") and "−" to Predication 2 ("it is not"), but for Predication 4, a clear notation is not evident. One attempt to introduce such a notation was by Matilal [19], who used "0" for Predication 4, but this does not accurately reflect the concept. An alternative notation was employed in the work of Haldane [6], who used 0, 1, and the Sanskrit symbol म्र (for Predications 2, 1, and 4, respectively) to indicate the seven standpoints, with the omission of the syllable *syāt*:

0, 1, म्र, 0 and 1, 0 and म्र, 1 and म्र, 0 and 1 and म्र.

However, Haldane [6] was connecting the conditional predication principle with finite arithmetic and computer architecture when he used म्र for Predication 4. Alternatively, Mardia [16] has used "+", "−", and "?" for Predications 1, 2, and 4, respectively. Predication 4 could otherwise be viewed in mathematical terms as akin to "0/0", but "?" is like seeing the amber light at a traffic signal, where there is ambiguity. In a survey, as another example, voting "Don't know" has a similar feeling, and the use of this notation leads to the formalisation of the seven predications as:

$$+, \quad -, \quad \pm, \quad ?, \quad +?, \quad -?, \quad \pm?$$

Importantly, the three main predications can be regarded as events A, B, and C, respectively, yielding the Venn diagram in Figure 10.2.

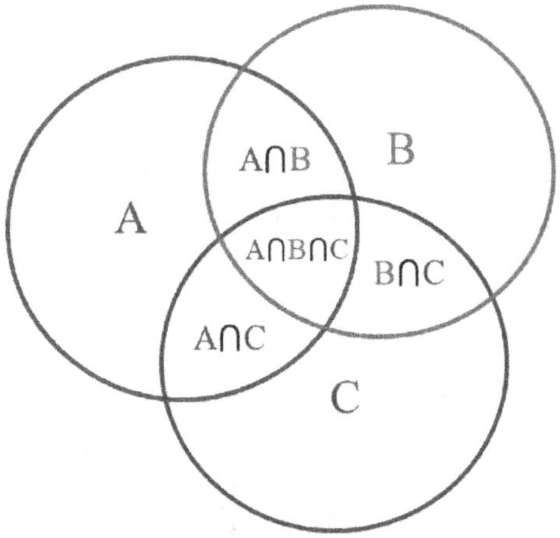

Figure 10.2. Venn diagram for *Saptabhaṅgī-naya*

In this Venn diagram, we have the three main Predications 1 (+), 2 (−), and 4 (?) as events A, B, and C, respectively, while the others are viewed as the intersections ($\bigcap$) of these events.

Additionally, Haldane [6] assigned the probabilities p, r, and q to Predications 1, 2, and 4, respectively, where $p + q + r = 1$, and considered a Markov chain for a particular learning experiment:

> "Now consider a subject who is shown a series of illuminated patches, some above his threshold of perception, some below it, and others very close to it, in a randomised series. We will suppose that he is in a steady state of sensory adaptation ... and that he is aware that his answers will sometimes be incorrect. At any given trial he will answer [p, q, or r]"

— or in our notation, "+", "−", or "?", respectively. Haldane [6] shows that unless one of p, q, or r is zero, the final predication will be the 7th, which we have denoted as "±?".

To summarise the statistical applications, we follow Mahalanobis [12] in noting that "1500 or 2000 years ago *syādvāda* seems to have given the logical background of statistical theory in a qualitative form."

10.3.5. *Conditional (non-absolute) holistic principle (Anekāntavāda)*

We now turn our attention to the core concept of Jain logic, *Anekāntavāda*, which we will refer to as the conditional holistic principle. Matilal [19] gives a first-order approximate translation of *Anekāntavāda* as "the theory of non-onesidedness" and further states that it "is, in fact, a unique contribution of the followers of Mahāvīra to the philosophic tradition of India."

Anekāntavāda is traditionally described by Jain logicians through the parable of six blind men (Mardia [16]):

> There are six blind men who want to know what kind of object an elephant is. Each touches a different part of the elephant. The one who touches a leg says "It is a pillar", the one who touches the tusk says "It is a pipe", the one who touches an ear says "This is a winnowing fan", and so on. Thus, each perspective differs, offering only a partial viewpoint (*nayavāda*), and so if we wish to understand what kind of object the elephant is, we must look at it from all sides.

In this way, we can see that the conditional predications of *Syādvāda*, applied to each entity, are like beads held together by the holistic principle of *Anekāntavāda* behaving like a thread. As L.C. Jain [7] summarised it, "*anekantavada* strives to incorporate the truth of all systems with its two organs: *nayavada* and the *syadvada*."

As this example implies, there is an art to sampling and interpretation that on the one hand reduces complex information to useful data through stratified sampling, and on the other delivers a recognisable approximation to truth (see Mardia and Cooper [17]). However, in this example the sampling is stratified but not totally random, as the six blind men are looking at different parts.

Mardia [14, 15] has indicated the relationship between this aspect of ancient Jain logic and the thinking of Karl Popper [22], who claimed that we cannot have absolutely true scientific laws. Mardia

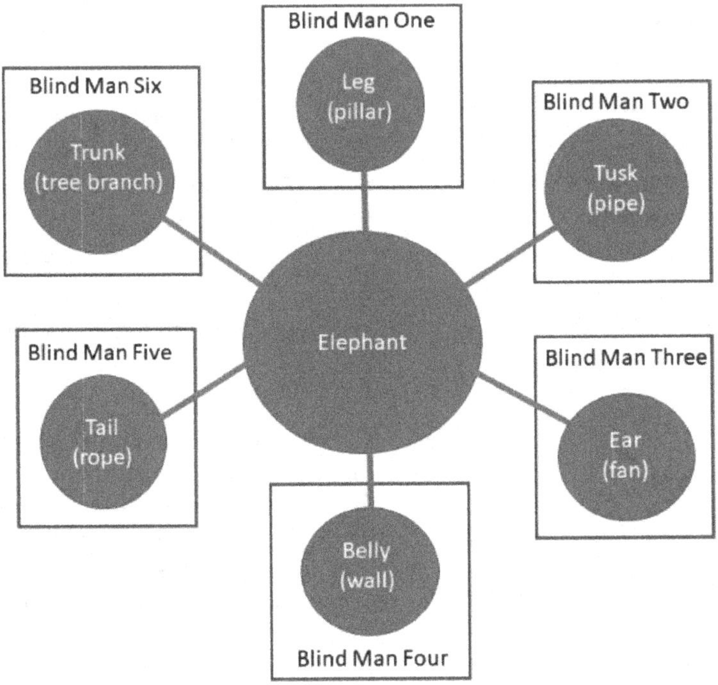

Figure 10.3. The blind men and the elephant, represented as a stratified sampling
The classic Jain parable can be framed as a stratified sampling, and illustrates how the conditional predications of *Syādvāda* are like beads held together with the holistic principle of *Anekāntavāda* behaving like a thread.

and Rankin [18] further remarked:

> "The Holistic Principle of Jain logic dovetails neatly with current trends in scientific and philosophical speculation that have been gathering momentum for the past half century. In *The Logic of Scientific Discovery* (1968), for example, Karl Popper expresses the view that there can be no 'absolutely true' scientific laws because our understanding is constantly developing and shifting, sometimes radically, sometimes at more subtle levels. This humanistic perspective is, in effect, a western counterpart to Anekantavada."

A similar line of thought was held by Einstein [3], who noted that one of the aims of science is "to reduce the connexions discovered to the smallest possible number of mutually independent conceptual elements. It is in this striving after the rational unification of the manifold that it encounters its greatest successes, even though it is precisely this attempt which causes it to run the greatest risk of falling a prey to illusions."

10.3.6. *Turing and 'enigmatic statistics'*

Returning to the conditional predications of *Syādvāda*, and the indeterminacy of Predication 4 in particular, we now give an example of indeterminacy inherent to the work of Turing on decoding the Enigma machine [24]:

> "Suppose a birder spotted 180 different species, many of which were represented by only one bird. Logically, other species must have been missed. A frequentist statistician would count those unseen species as zero, as if they could never be found. Turing, by contrast, assigned them a tiny non-zero probability, thereby factoring in that rare letter groupings might not be present in his current collection of intercepted messages but could appear in a larger sample."

And so, from one standpoint, there are no other species; from another, there are other species; and from Turing's standpoint, it is indeterminate.

Revisiting the Bayesian aspects of Section 10.3.2, we adapt an example that Turing [28] gave of determining life expectancy by drawing upon prior information, in which he employed a Bayesian analysis of conditional probability: "The *probability* of an event on certain evidence is the proportion of cases in which that event may be expected to happen given that evidence."

For instance, if it is known that 70% of all men live beyond the age of 45, then knowing of Turing only that "he is a man", we can say that the probability of Turing living past the age of 45 was 0.7. Suppose, however, we know that "Turing is now of age 30" (c. 1942), then the probability will be quite different, say 0.8, on the basis that 80% of men of age 30 will live past 45.

We know, of course, that Turing's life was tragically cut short at the age of 41, but even by then his legacy in mathematics was assured.

And while much is known of his contributions to cryptography, computer science, and artificial intelligence, his work in statistics is also significant and not to be overlooked. Indeed, the example adapted above came during his wartime efforts at Bletchley Park, where he used various statistical methods in his work on decoding the Enigma machine [24]. But while this aspect of Turing's work and his techniques remain less widely known, we might refer to it as 'enigmatic statistics' (see Mardia and Cooper [17]).

10.4. Conclusion

By providing a comprehensive analysis of the five-fold Jain syllogism (*Pañcāvayavavākya*) together with the historical perspective, we have unified the literature and highlighted how ancient Jain logic applies to modern statistical thinking. In particular, we have shown how combining the five-fold Jain syllogism with the conditional predication principle *Syādvāda* (by adding *Syāt* to the Jain syllogism) results in a Bayesian approach. We have also presented a plausible notation for the *Syādvāda* system and used a Venn diagram to illustrate the concept, thus making these Jain ideas more accessible within the terminology of modern logic. Connection with the work of Turing is also illuminating, as his less celebrated statistical techniques merit deeper investigation.

As ancient Jain logic has recognised for millennia but only came to light in modern mathematics in the previous century (when Gödel's incompleteness theorems [5] demonstrated that the truth of some mathematical propositions are inherently undecidable), a firm grasp of the absolute truth is beyond our reach. Instead, we have endeavoured to show, following Mardia and Cooper [17], how "the role of mistakes, uncertainty, interaction, 'common sense', and [intuition and Bayesian methods] all point to a world in which logic and statistical methods work together, in complementary ways," much in the way that *Anekāntavāda*, *Syādvāda*, and *Pañcāvayavavākya* may be employed to reveal the multifaceted nature of truth.

We conclude with the following broad summarisation of the Jain worldview from Mahalanobis [12]:

"Finally, I should draw attention to the realist and pluralist views of Jain philosophy and the continuing emphasis on the multiform and infinitely diversified aspects of reality which amounts to the acceptance of an 'open' view of the universe with scope for unending change and discovery."

Acknowledgment

We wish to thank Acharya Nandighoshsuriji for his help with scriptural insights and to thank Professor Arni S.R. Srinivasa Rao of Augusta University for his helpful suggestions. We also wish to thank (posthumously) the late S. Barry Cooper, Professor of Mathematical Logic at the University of Leeds, for the discussion on the work of Alan Turing described here. Further, we would like to thank the anonymous reviewers for their comments, which have improved the paper considerably.

References

[1] N. N. Bhattacharyya, *Jain Philosophy: Historical Outline.* Munshiram Manoharlal Publishers, 1999, pp. 138–141. Available at: https://archive.org/details/JainaPhilosophyHistoricalOutlineNarendraNathBhattacharyyaMRML.

[2] A. Bolstad, B. J. Carlin, H. S. Stern, D. B. Dunson, A. Vehtari, and D. B. Rubin, *Bayesian Data Analysis.* 3rd ed. CRC Press, 2013. Available at: http://www.stat.columbia.edu/~gelman/book/BDA3.pdf.

[3] A. Einstein, Science and religion, *Nature*, **146** (1940), 605–607.

[4] M. J. Evans and J. S. Rosenthal, *Probability and Statistics: The Science of Uncertainty.* 2nd ed. Freeman (2009), 7–10. Available at: http://www.utstat.toronto.edu/mikevans/jeffrosenthal/book.pdf.

[5] K. Gödel, *Über formal unentscheidbare Sätze der Principia Mathematica und verwandter Systeme I* (*On Formally Undecidable Propositions of Principia Mathematica and Related Systems*, B. Meltzer, trans.). Dover Publications, 1992.

[6] J. B. S. Haldane, The syadvada system of predication, *Sankhyā*, **18**(1/2) (1957) 195–200.

[7] L. C. Jain, Perspective of system-theoretic technique in Jaina school of mathematics between 1400–1800 A.D., *Jain Journal.* Jain Bhawan Publication, **XIII**(2) (1978) 49–66.

[8] L. C. Jain and C. K. Jain, *The Jaina ulterior motive of mathematical philosophy.* Jain Prachy Vidhayya, (1987) 49–59.

[9] M. K. Jain, Logic of evidence-based inference propositions, *Current Science*, **100**(11) (2011) 1663–1672.

[10] J. L. Jaini, *Outlines of Jainism*. Cambridge University Press (1916) 117–118.

[11] D. S. Kothari, The complementarity principle and eastern philosophy, *Neils Bohr Centenary Volume* (French, A.P. and Kennedy, P.J., eds). Cambridge: Harvard University Press (1985) 325–331.

[12] P. C. Mahalanobis, The foundations of statistics, *Dialectica*, **8**(2) (1954), 95–111. Reprinted in *Sankhyā: The Indian Journal of Statistics (1933–1960)*, **18**(1/2) (1957) 183–194.

[13] Yuvacarya Mahāprajña, *Jaina Nyāya Kā Vikāsa* (*New Dimensions in Jaina Logic*, N. Tatia, trans.). Jaina Vishva Bharati (1984) 123–125.

[14] K. V. Mardia, Jain logic and statistical concepts, *Jain Antiquary and Jaina Siddhanta Bhaskar*. Oriental Research Institute, **27**(1975), 33–37.

[15] K. V. Mardia, Discussion to "Probability, Statistics and Theology", by D. J. Bartholomew, *Journal of the Royal Statistical Society, Series A*, **151**(1) (1988) 166–167.

[16] K. V. Mardia, *The Scientific Foundations of Jainism*. Motilal Banarsidass (1990) 94–99.

[17] K. V. Mardia and S. B. Cooper, Alan Turing and enigmatic statistics, *The Once and Future Turing* (S. B. Cooper and A. Hodges, eds). Cambridge University Press (2016) 78–89.

[18] K. V. Mardia and A. D. Rankin, *Living Jainism: An Ethical Science*. Mantra Books (2013) 56.

[19] B. K. Matilal, *The Central Philosophy of Jainism (Anekānta-vāda)*. L.D. Institute of Indology, Ahmedabad (1981) 1–6. Available at: https://archive.org/details/CentralPhilosophyOfJainismAnekatvaVadaBimalaKrishnaMatilalL.D.InstituteOfIndology/.

[20] S. Mookerjee, *The Jaina Philosophy of Non-Absolutism*. Bharati Jaina Parisat (1944) 5, 78, 115.

[21] S. Mookerjee and N. Tatia, *Hemacandra's Pramāṇa-mīmāṃsā*. Tara Book Agency (1984) 132–134.

[22] K. R. Popper, *The Logic of Scientific Discovery*. 2nd ed. Hutchinson & Co., London (1968) 280.

[23] G. N. Ramachandran, Syad Naya System (SNS) — a new formulation of sentential logic and its isomorphism with Boolean algebra of genus 2. *Current Science*, **51**(13) (1982) 625–636.

[24] A. Robinson, Statistics: Known unknowns, *Nature*, **475** (2011) 450–451.

[25] J. M. Rogers and M. K. Jain, Inference and successful behavior, *The Quarterly Review of Biology*, **68**(4) (1993) 387–397.
[26] S. Shastri, *Tark Sangraha of Anambhatt* (original text and Gujarati commentary). Shree Umra Murtipujak Jain Sangh, Surat (2016) 163.
[27] S. K. Thompson, *Sampling*. 3rd ed. Wiley (2012) 141–156.
[28] A. M. Turing, The applications of probability to cryptography, *The National Archives*, HW 25/37, 1941/1942a. Available at: https://discovery.nationalarchives.gov.uk/details/r/C11510465.
[29] S. C. Vidyabhusana, *A History of Indian Logic: Ancient, Mediaeval and Modern Schools*. Motilal Banarsidass, 1971, pp. 161–162, 166–167, 221, 497–499. Available at: http://library.bjp.org/jspui/bitstream/123456789/846/1/History-Of-Indian%20Logic-1921-Philosophy.pdf.

https://doi.org/10.1142/9789811255502_0011

Article 11

Zero in the Early Śvetāmbara Jaina Works

Achary Vijay Nandighoshsuri

C/o Research Institute of Scientific Secrets from Indian Oriental Scriptures-RISSIOS
Ahmedabad 380001, India

There are references to zero in Jaina scriptures, āgamas like *Jambūdvīpaprajñapti*, *Jīvābhigama*, *Bhagavatī Sūtra* etc., which were originally composed in the millennium BCE and preserved the in oral tradition up to the 5th century CE. On the basis of language, grammatical forms and related material Professor Hermann Jacobi concluded that the present version of Jaina Śvetāmbara Aṅgas must have been compiled towards the end of the 4th century BCE and at the beginning of the 3rd century BCE. In this article we discuss instances of occurrences of zero in the Jaina literature.

Keywords: Zero, Manuscript, Number, *Aṅgas*, *Upāṅgas*, *Āgama*, Commentary, *Yojana*, Area, *Śîrṣaprahelikā*.

Mathematics Subject Classification 2020: 01-06, 01A32, 01A99.

According to the Jaina chronological history, 980 years after the Nirvāṇa of Lord Mahāvīrasvāmî, until Devardhi Gaṇi Kṣmāśramaṇa (454 CE), all of the Jaina āgamas came through oral tradition; thus there are no documented Jaina āgamas prior to 454 CE, according to śvetāmbara tradition. Under the pious leadership of Devardhi Gaṇi Kṣmāśramaṇa, 500 ācāryas wrote the āgamas on palm-leaves. This was a revolutionary step in the Jaina āgamic literature. In this āgamic literature there are 12 aṅgas and 12 upāṅgas that are primary. However, the 12th agama named *Dâṣtivāda* was forgotten, at an early time, so at present we have only 11 aṅgas and 12 upāṅgas, which were compiled by Devardhi Gaṇi *et al.* [1].

The *Jambūdvīpaprajñapti* is the sixth upāṅga, which was compiled during this Valabhī Vācanā (454 CE). The *Jambūdvīpaprajñapti, Sūryaprajñapti, Candraprajñapti, Dvīpasāgaraprajñapti* etc. Āgamas belong to the *Gaṇitānuyoga* [2]. In these āgamas, a large portion is occupied by discussion of mathematical operations concerning planar Geometry on a large scale like that of Jambūdvīpa having a diameter of 1,00,000 yojanas. In these āgamas all mathematical calculations were performed using the decimal system. We find mathematical operations, not only in *Jambūdvīpaprajñapti, Sūryaprajñapti, Candraprajñapti and Dvīpasāgaraprajñapti* but also in other āgamas as well [3].

In the editorial notes contained in the *Jambūdvīpaprajñapti* [4], Ācārya Śrī Munichandrasuri has mentioned that under the pious leadership of Devardhi Gaṇi *et al.* when all the āgamas were written on palm-leaf, the original scriptures would have been shortened, because with time a large portion of the original scripture would have been lost or forgotten and this lost portion would have been substituted from the *Jīvābhigama* by Upādhyāya Śānticandra [5]. In such a way a large portion of the original scripture was lost during oral tradition, i.e. before 454 CE.

On this āgama, *Jambūdvīpaprajñapti*, Ācārya Malaygirijī has also composed a commentary in the first half of twelfth century CE [6]. Another commentary on *Jambūdvīpaprajñapti* was composed by Muni Śrī Puṇyasāgara of the Kharataragaccha in Vikrama Saṃvat 1645 (1589 CE), while a third commentary was composed by Upādhyāya Shrī Śānticandra in Vikrama Saṃvat 1660 (1604 CE). In the original text, namely sūtras which are in Ardhamāgadhī language, Lord Mahāvīra had given a description of Jambūdvīpa as a circular island (dvīpa). Not only had he given statistical data, but also a procedure to calculate its circumference and area as follows:

एगं जोयणसयसहस्सं आयाम,विक्खंभेणं तिण्णि जोयण सयसहस्साइं सोलस।सहस्साइं दोण्णि य सत्तावीसे जोयणसए तिण्णि य कोसेअट्ठावीसं धणुसयं तेरस य अंगुलाइं अद्धंगुलं च किंचिविसेसाहियं परिक्खेवेणं पण्णत्ते ।।

Thus, the circumference of Jambūdvīpa, whose diameter is 100,000 yojana, is given to be 316227 yojana, 3 *kose*, 128 *dhanu*, 13 and half aṅgula and a little over; the latter are names of different successively finer units. This is calculated as the square root of 100,000,000,000. They widely used $\sqrt{10}$ for the ratio of the circumference to the diameter.

To calculate the area of Jambūdvīpa, it is mentioned that the circumference should be multiplied by a fourth of the diameter, or equivalently half of the radius; the relevant text is as follows:-

विक्खंभवग्गदहगुणकरणी वट्टस्स परिरओ होइ । विक्खंभपायगुणिओ परिरओ तस्स गणियपयं ।

Here परिरओ means circumference, गणियपय means area, विक्खंभ means diameter, करणी means square-root and वट्टस्स means of the (specific) circle.

There is also a procedure described to calculate lengths of arcs (cāpa) of circular objects like Jambūdvīpa.

The calculation of circumference as above shows that our ancient Jaina ācāryas were very much familiar with use of the decimal system, in the millennium BCE. Without the use of zero and the decimal system calculation with such large numbers is unimaginable. Also, this is just one, and a relatively simpler calculation, compared to other calculations in Jaina literature.

A conference of well-versed learned, scholars and prominent Jaina ācāryas was held in 300 BCE at Pāṭaliputra (modern Paṭanā), presided over by Ārya Sthulibhadra, who was the chief disciple of Ārya Bhadrabāhu (the sixth ācārya in the lineage of Lord Mahāvîra, who is held in high respect by all sects of Jainas). In the same spirit a 2nd conference was held in 300 CE, at two places simultaneously: at Mathurā, presided over by ācārya Skandila, and at Valabhī, presided over by Nāgārjuna. However, the oral tradition continued. Then again a third conference (the last of the original ones) was summoned by Devardhi Gaṇi Kṣmāśramaṇa at Valabhī in 454, at which, as mentioned above, the āgamas were put down on record in the palm-leaf medium. On the basis of the language, grammatical forms and related material Professor Hermann Jacobi concluded that these aṅgas must have been compiled towards the end of the 4th century BCE and the beginning of the 3rd century BCE.

Admittedly there is no tangible evidence of zero in written form in the *Anuyogadvāra Sūtra* and the *Bhagavatī Sūtra* at the time of compiling the āgamas in 454 CE [7]. The word śūnya was used for zero. *Anuyogadvāra Sūtra* stated the maximum number of human bodies in this *Aḍhīådvīpa*. It is described to be greater than *tiyamalapaya* (a 24-digit number) and less than *cauyamalapaya* (a 32-digit number). It turns out that there are 29 digits in this number. It is the product of $2^{32} \times 2^{64} = 2^{96}$, which is equal to 792,281,625,142,643,375,935,439,5**0**3,36. Notably, here in 1000's place it has śūnya [8].

It is true that this figure of 29 digit is not given in the original *Anuyogadvāra Sūtra*; however, it is available in commentaries.

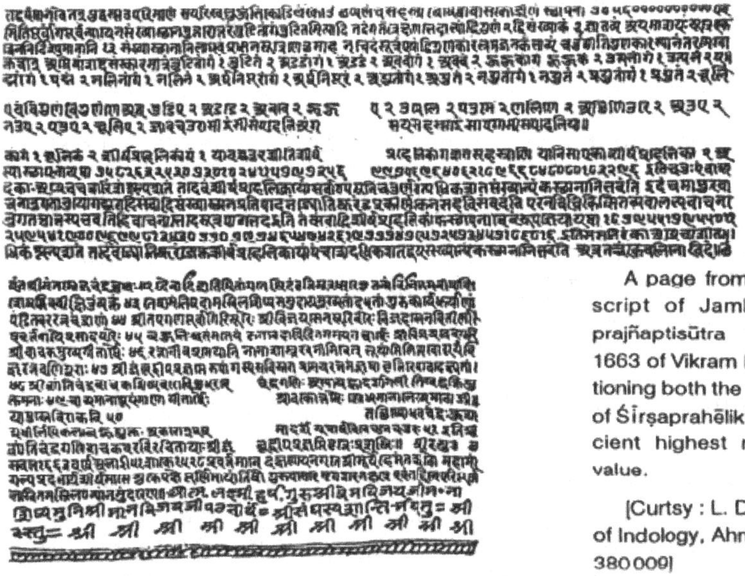

A page from a manuscript of Jambūdvīpa prajñaptisūtra written in 1663 of Vikram Era, mentioning both the variations of Śīrṣaprahēlikā, the ancient highest numerical value.

[Curtsy : L. D. Institute of Indology, Ahmedabad 380009]

The largest possible figure is mentioned in the two systems: one is the decimal system and the second with base of 84,00,000, which has been a unique feature of Jaina āgamas. This largest figure is named as *Śirṣaprahelikā* and, according to Māthurī vācanā it contains 194 digits, while according to Valabhī vācanā it contains 250 digits. From these two versions of Śirṣaprahelikā, in the first version there are 54 different digits followed by 140 zeros and in the second version there are 70 different digits followed by 180 zeros see [9].

According to Māthurī vācaṇā: $(84{,}00{,}000)^{28} = 758, 263, 253, 073, 010, 241, 157, 973, 569, 975, 696, 406, 218, 966, 848, 080, 183, 296 \times 10^{140}$

According to Valabhī vācanā: $(84{,}00{,}000)^{36} = 187, 955, 179, 550, 112, 595, 419, 009, 699, 813, 430, 770, 797, 465, 494, 261, 977, 747, 657, 257, 345, 718, 6816 \times 10^{180}$

Such a large figure without any mistakes is very well preserved for close to one thousand years from the Nirvāṇa of Lord Mahāvīra in oral tradition and then until today in manuscripts. There are a large number of zeros in this largest figure. It shows that this oral tradition proves that zero was used in Jaina scriptures since ancient times.

Thus, to summarize, in the Jaina tradition zero was used in the first millennium BCE, though unfortunately we do not have any scriptural proof for this. However, according to tradition what we have at present in manuscript form was not new, but was originally copied from former manuscripts, and ultimately it was orally queried by Jambūsvāmî or Sudharmā svāmî and they received this information from Lord Mahāvīra.

Before concluding we may also mention that according to the Jaina tradition mentioned in many more ancient Jain canonical manuscripts, mathematics had been taught by Lord Ādinātha to Sundarī [10], [11].

References

[1] Ārya Bhadrabāhu, Commentary by Vinayavijaya, *Kalpasūtra, Part-6,* (Śrī Mātungā Jain Svetāmbara Mūrtipūjaka Tapāgaccha Sangh, Mātungā, Bombay, 1975), Folio 360, 361, 362.

[2] Sarju Tiwari, Mathematics in History, Culture, Philosophy and Science (Mittal Publication, 1992), 81 Also see Gaitåný́yoga, ed. by Muni Kanhaiyalal Kamal, Sanderao (Raj.), 1986.

[3] Malaygirijī, *Commentary on NandiSūtra,* Agamodaya Samiti, Surat, 238.

[4] Ārya Jambýsvāmi Commentary by Upādhyāya Śānticandra, *JambuddīvpannattiSuttá* (Edited by Ācārya Śrī Munichandrasuri and Published by Omkar suri Gyanmandir, Surat, 2017 AD) 8.

[5] Ibid, P. 9, 10, 24, 25, 26.

[6] Walter Schubring: Doctrine of Jainas, (Motilal Banarasidas, Delhi, 1968) 77, 78, 81.

[7] Ārya Jambūsvāmi *Anuyogadvār Sūtra* Vol. 1, Anuyogacandrikā commentary and Vol. 2 Prameyacandrikā commentary by Ghisulāljī (Akhil Bhārtīya Svetāmbar Sthānakvāsī Jain Sāstroddhāra Samiti, Rajkot, 1968).

[8] R. S. Shah, Pune: *Mathematical Ideas in Bhagavatī Sūtra (BS)*, Gaṇita Bhāratī (Delhi) **30**(1), PP. 1–25 2008.

[9] Munishri Nandighoshvijayji Gaṇi, *Scientific Secrets of Jainism* (Research Institute of Scientific Secrets from Indian Oriental Scriptures-RISSIOS, Ahmedabad-380001, India, 2001) 103.

[10] अत्र लिखितं हंसलिप्याद्यष्टादशलिपिविधानं, तच्च भगवता दक्षिणकरेण ब्राह्म्या उपदिष्टम्, गणितं तु – एकम्, दशम्, शतम्, सहस्रम्, अयुतम्, लक्षम्, प्रयुतम् कोटिः, अर्बुदम्, अब्जम्, खर्वम्, निखर्वम्, महापद्मम्, शङ्कुः, जलधि, अन्त्यम्, मध्यम्, परार्धम् चेति यथाक्रमं दशगुणम् इत्यादि संख्यानं सुन्दर्याः वामकरेण, काष्ठकर्मादिरूपम॒ कर्म भरतस्य, पुरुषादिलक्षणं च बाहुबलिनं उपदिष्टम् ।।

Ārya Bhadrabāhu, Commentary by Vinayavijaya, *Kalpasūtra, Part-7,* (Śrī Mātungā Jain Śvetāmbar Mūrtipūjak Tapāgaccha Sangh, Mātungā, Bombay, 1975), Folio 443.

[11] अष्टादश लिपिर्ब्राह्म्या, अपसव्येन पाणिना । दर्शयामास सव्येन, सुन्दर्या गणितं पुनः ।।963।।

Hemacandrācārya, KalikālSarvajña, *Triṣaṣṭhiśalākāpuruṣacaritramahākāvya Parva-1, Sarga-2, Stanza-963,* Editor: Munirāja Śrī Caraṇavijayjī Mahārāj (K. S. Hemacandrācārya Navam Janmaśatābdi Smriti Sanskār Śikṣaṇanidhi, Ahmedabad, 2015) 57.

https://doi.org/10.1142/9789811255502_0012

Article 12

Jaina Mathematical Sources since the Eighth Century

Anupam Jain

Centre for Studies of Ancient Indian Mathematics
D. A. University, Indore 452001, India
anupamjain3@rediffmail.com

In this paper we present a survey of the non-canonical Jaina sources containing mathematical ideas, from 8th century onward, that are presently available. The sources are categorized in two parts: Unpublished texts, and published texts.

Keywords: Jaina mathematical manuscripts, Published Jaina mathematical texts.

Mathematics Subject Classification 2020: 01-06, 01A32, 01A99.

12.1. Introduction

It is well-known that Jaina scholars have engaged themselves in mathematics since the very early times. We find that even in the works that are largely philosophical, including the canonical works, contain mathematical ideas. In the later period, gradually more and more works primarily devoted to mathematics were also composed by Jaina scholars. The study of these, however, has not received adequate attention of modern scholarship. Lack of awareness of the sources available seems to have been one of the factors in this respect. There is now some movement, albeit feeble, towards putting together information from Jaina works so as to make it accessible to scholars and other interested readers, in which the present author has been involved. This article is an endeavor to put together the inputs gathered in the form of a profile of the current state concerning accessibility of the Jaina works with mathematical content.

In the last 100 years or so, many Jaina texts were discovered and published with Hindi/English translations. During the search through Jaina bhaṇḍārs (repositories of manuscripts attached with Jaina places of worship) many manuscripts previously unknown to the scholarly community were discovered. These manuscripts throw light on considerable contributions by the Jaina ācāryas (teachers - monks) in the field of Mathematics and Astronomy.

The mathematical topics involved in the Jaina works played a role as tools for the ācāryas (teachers) to illustrate the philosophical and other points discussed in their works. The motivation for this involves the following aspects:

1. To formulate and describe various cosmographical details of the three-fold universe, such as the sizes of different mountains, rivers and parts of our Jambūdvīpa (the earth?).
2. To explain different types and sub-types of "*karmas*" (our deeds, reckoned in spiritual terms) and the operations of uprising, binding and shedding off *karmas*, and the net effect of the infinitude of their combinations.
3. To calculate the auspicious place and time for performance of religious ceremonies, like *dīkṣā* (initiation) and *Pratiṣṭhā* (consecration).
4. To train the householders in basic mathematics (*Laukika Gaṇita*), required in daily life and in construction of temples and houses.
5. To explain Jaina system of logic used for arguments and reconciliation.

To explain the mathematical contents of Jaina *āgamas* many commentaries were written by Jaina scholars, including mathematical texts. Unfortunately, many of them seem to have been lost.

Nevertheless, we find many mathematical texts in the surviving Jaina literature. The aim of this article is to give a brief exposition of the status in respect of the works known so far.

12.2. Unpublished Non-Canonical Jaina Mathematical Texts

In the following, the details of the manuscript are given in the following order, along with an image pertaining to the manuscript: Name of the Manuscript, Other Names, Author's Name, Period of Manuscript, Subscribing Period, Source, Manuscript No, Folios, Subject, Language, Script, Status (complete, incomplete, etc.). Other known manuscripts (if any). Information about photocopies of the individual manuscripts acquired by the author is also included.

1. *Gaṇita Sāra* (*Triśatikā*) by Śrīdharācārya (750 C.E)

Gaṇita-Sāra, *Triśatikā or Pātī-gaṇita-sāra*, Ācārya Śrīdhara, 8th c. CE, X, available at (i) Jaina Matha, **Moodabidrī**, 781, 26, Mathematics, Saṃskṛta, Kānarī, (Complete), (ii) British Museum Catalogue No. 5205, (iii) India Office London, Pātya Gaṇitasāra, No. 6317, Triśati, and No. 2787, 2788, 2789, 2790, (iv) Mohanlal Library, Mumbai, No. 156. (There are also copies elsewhere).

Several manuscripts of *Triśatikā* have been collected at the Department of Mathematics, University of Lucknow, Lucknow, from various places. The published version[1] of *Triśatikā* is incomplete.

It has only 172 Ślokas (verses), whereas as the title of the manuscript indicates there should be 300 Ślokas. Of the 172 Ślokas 65 describe various formulas and the remaining 107 Ślokas are devoted to examples. I have acquired a copy of *Gaṇitasāra* (*Triśatikā*) from Moodabidrī and one version of *Triśatikā* prepared by Usha Asthana (1960) in her thesis [3], based on several available transcripts. I have also a copy from Scindhia Oriental Manuscript Library-Ujjain (Mss No. 7167). This version is in 22 Folios and has about 300 Ślokas. The name in the *Praśasti* is *Triśatī* and the name of the scriber is Ramacandra.

The manuscript of Moodbidrī has a *maṅgalācaraṇa.*

नत्वा जिनं स्वविरचित पाट्या, गणितस्य सारमुद्घृत्य ।

लोक व्यवहाराय प्रवक्ष्यति, श्रीधराचार्य: । 01 ।[2]

Śrīdharācārya, bowing to (Lord) Jina, tells the essence of Mathematics as extracted from the pātī composed by himself, for the use of the people. (pātī here refers to the larger work of the author.)

An example bearing the names of *Tīrthaṅkaras* (the original spiritual teachers of the Jaina tradition) is found in a manuscript collected by Prof. A.N. Singh (Lucknow).

वृषभे सम्भवे पंच सप्तोन्दुरूत दीशयोः

विमलेऽर्द्धम् चतुर्द्वारकमणैकं च दृश्यते । 98 ।[3]

It may be noted here that *Vṛṣabha*, *Sambhavanātha* and *Vimalanātha* are names of the first, third and 13th Tīrthaṅkaras, respectively, in the Jaina tradition. Due to lack of awareness of this detail about the Jaina tradition, in her thesis [3] Usha Asthana has incorrectly translated *Vṛṣabha* as nandi, *Sambhava* as Ganesh and *Vimale* as "in the process of cleaning".

A new critical edition of *Triśatikā* ought to be brought out urgently, based on various versions as discussed above.

2. *Jyotirjñānavidhi* of Śrīdharācārya (750 C.E.)

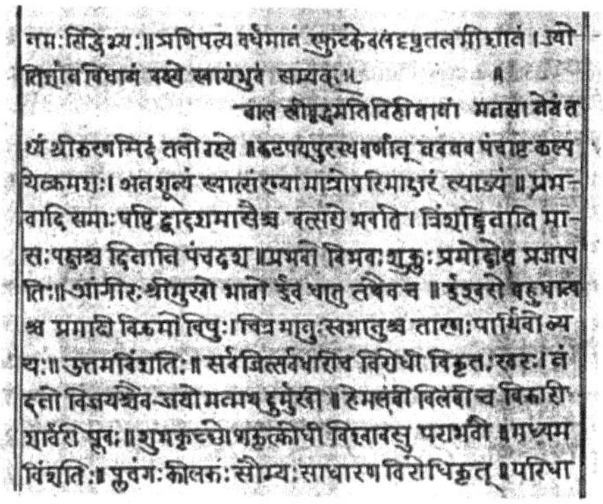

Jyotirjñānavidhi, Śrīkaraṇa, Ācārya Śrīdhara, 799 CE, X, Śrī Ailaka Pannālāla Digambara Jaina Sarasvatī Bhaṇḍāra, **Ujjain**, 661, 16, Astronomy, Saṃskṛta, Devanāgarī, (Complete).

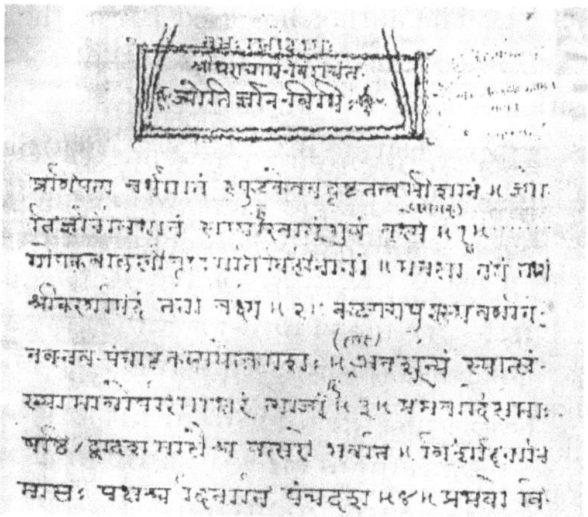

Jyotirjñānavidhi, Śrīkaraṇa, Ācārya Śrīdhara, Vīravāṇī Vilāsa, Jaina Siddhānta bhavana, Moodabidrī, Mss. No. 47, Folio-9 & its transcript at Jaina Siddhānta Bhavana, **Arrah**, Jha/137/01, 16. Saṃskṛta, Devanāgarī, (Complete).

So far, the text of it has only been published by the Kundakunda Jñānapīṭh, Indore, and moreover that also requires many modifications.

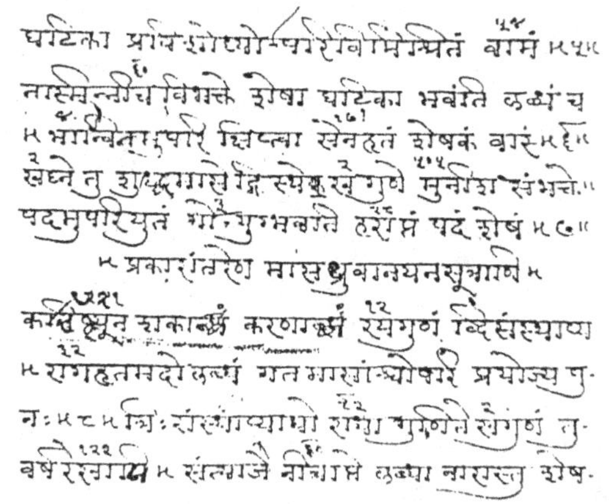

From internal evidence in *Jyotirjñānavidhi* it is clear that it is composed in 799 CE. The author has used 721 as the śaka year and 720 as the previous śaka year, in chapter-2 at 2 places [4]. Hence the date of composition is śaka 721 which is equivalent to the 799 CE. The arguments of B.B. Datta, R.C. Gupta, N.C. Shastri, Mamta Singhal, Shaifali Jain and others also support this.

With the help of two Manuscripts found in Ujjain and Arrah we have prepared [5] a list of the titles of the 10 Chapters to be the following:-

1. Saṃjñādhikāra
2. Dhruvādhikāra
3. Tithyādhikāra
4. Saṃkrānti-Ṛtvahorātri pramāṇādhikāra
5. Grahanilayādhikāra
6. Grahayuddhādhikāra
7. Grahaṇādhikāra
8. Lagnaprakaraṇa
9. Gaṇitādhikāra
10. Muhūrtādhikāra

Clearly it is a book on Mathematical Astronomy. It should be published together with a translation, as it is a book of first rate importance.

3. *Lokānuyoga* by Jinasena (8th c. CE)

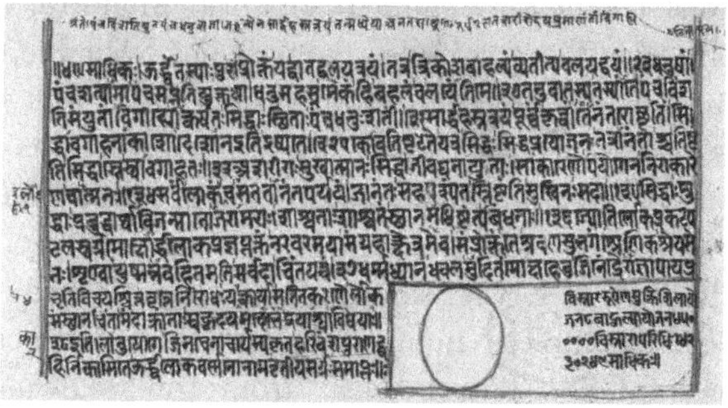

Lokānuyoga, 4th, 5th, 6th ch. of *Harivaṃśa purāṇa*, Jinaṣena-I, 783 C.E, X, (i) Śrī Ailaka Pannālāla Digambara Jaina Bhaṇḍāra, Ujjain, 43, 73, Structure of Cosmos, Saṃskṛta, Devanāgarī, (Complete), (ii) Śri Mūlasaṃgha Candranātha Svāmī Balātkāra-gaṇa Digambara Jaina Mandira, Karanja (Lad) (Vasim) 117/612, (iii) Anekānta Śodhapīṭha, Kumbhoja-Bāhubalī, 10/24.

The published version of *Harivaṃśapurāṇa* [6] contains the ślokas of *Lokānuyoga* but this copy is more detailed and contains illustrations. It is thus more useful from a mathematical point of view.

4. Ṣaṭtrinśikā of Mādhavacandra Traividya (11th c. CE)

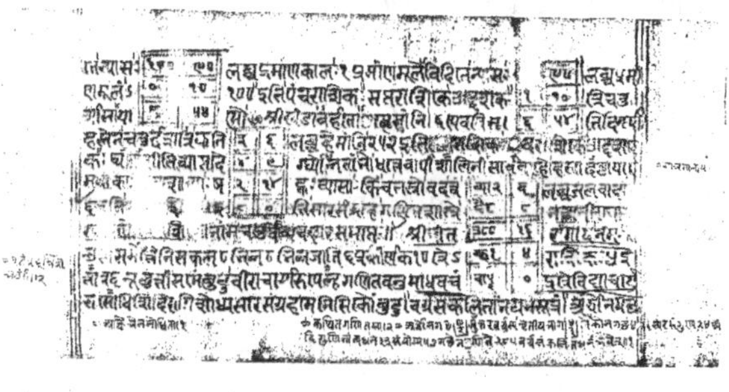

Ṣaṭtrinśikā, *Ṣaṭtrinśatikā* or *Chattīsī Tīkā*, Mādhavacandra Traividya, 11th c. CE, 1664 V.S. [1607 CE], (i) Śrī Digambara Jaina Temple, Tholion, Jaipur, 468 (Vesthana No 465), Patra 38–45, Mathematics, Saṃskṛta, Devanāgarī, (Complete), (ii) Digambara Jaina Blātkāragaṇa Mandira, Karanja, Mss. No. 63 & 65, Folio-49 & 53, (iii) Shri Digambara Jaina Bīsapanthī Mandira, Mandi Nal, Udaipur, Folio-53. (There are also more copies available elsewhere).

Mādhavacandra Traividya is a commentator of *Trilokasāra* (*Tiloyasāra*) of Nemicandra (981 CE) *Ṣaṭtrinśikā* was written by him based on *Gaṇita-Sāra Saṃgraha* (GSS) of Mahāvīrācārāya (850 CE) There was some confusion about the authorship of this work. The four chapters are the same as in GSS. Hence everywhere it is listed as incomplete manuscript of GSS, but actually it is an extension of GSS. This is clear from the following colophon of Mādhavacandra Traividya in his commentary on *Trilokasāra* (*Tiloyasāra*) of Nemicandra (981 CE).

श्री वीतरागाय नमः |6| छत्तीसमेतेन सकल 8, भिन्न 8, भिन्न जाति 6, प्रकीर्णक 10, त्रैराशिक - 4 इंता 36 नू छत्तीस मे बुदुवीराचार्यरूपेल्हगणितवनु माधवचंदत्रैविद्याचार्यारू शोध सिदरागि शोध्यसार संग्रहमेनिसिकोंबुदु |

The Chapters of GSS are the following:-

1. Sanjñādhikāra-Parikarma Vyavahāra
2. Kalāsavarṇa Vyavahāra
3. Prakīrṇaka Vyavahāra
4. Trairāśika Vyavahāra

13 new rules, related to series, geometry and typical problems have been added by Mādhavacandra Trividya.

The New Formulae are the following:-

1. Varga Saṃkalitānayana Sūtraṃ.
2. Ghana Saṃkalitānayana Sūtraṃ.
3. Ekvārādisaṃkalitā Ghānanayan Sūtraṃ.
4. Sarva Ghananayane Sūtra Dvayaṃ
5. Uttrottaracaya Bhava Saṃkalitāghanānayana Sūtraṃ.
6. Ubhyāntādāgata Puruṣadvaya Saṃyogānayana Sūtraṃ.

7. Vaṇikkara Sthita Ghanānayana Sūtraṃ.
8. Samudra Madhye 1-2-3.
9. Chedośaśesa Jāto karaṇa Sūtraṃ.
10. Karaṇa Sūtra Trayaṃ.
11. Guṇaguṇya Mīśre Sātiguṇaguṇyānayana sūtra.
12. Bāhu karṇānayana Sūtraṃ.
13. Vyāsādyānayana Sūtraṃ.

Magic squares in which the row, columns and diagonals sums are 34 are very common in Jaina tradition. Such a magic square is also found in this manuscript.

9	16	2	7
6	3	13	12
15	10	8	1
4	5	11	14

In fact, several manuscripts which have presently been documented as incomplete manuscripts of GSS could be copies of *Saṭtrinśikā*. Its publication is also required.

5. *Trailokya Dīpaka* of Paṇḍita Vāmadeva (14th c. CE)

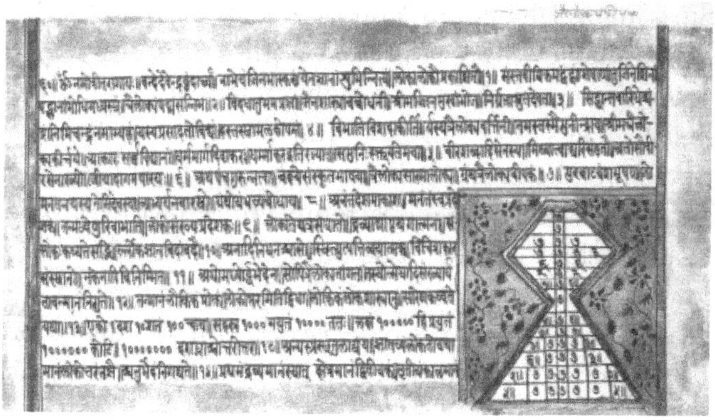

Trailokya Dīpaka, X, Pt. Vāmadeva (Prior to 1379 CE), 1988 V.S., (i) Amara Granthālaya, Indore, 293, 62, Structure of Cosmos, Saṃskṛta, Devanāgarī, (Complete), (ii) Anekānta Vacanālaya, Chanderi, Mss. No 554 Folios-83, (iii) Ācārya Kundakunda Hastalikhita Śāśtra Bhaṃdāra, Khajuraho, Mss. No. 64, Folios-20, (iv) Ailaka Pannālāla Sarasvatī Bhaṃdāra, Ujjain, Mss No 485, Folios-89.

This is a text about the three-fold universe, and its contents are almost the same as in *Tiloyapaṇṇattī* of Yativṛṣabha (176 CE) and *Tiloyasāra* of Nemicandra (981 CE). Study of the above manuscript may yield new information about these other manuscripts as well.

6. *Uttara Chattīsī Tīkā* by Bhattāraka Sumatikīrti (16th C. CE)

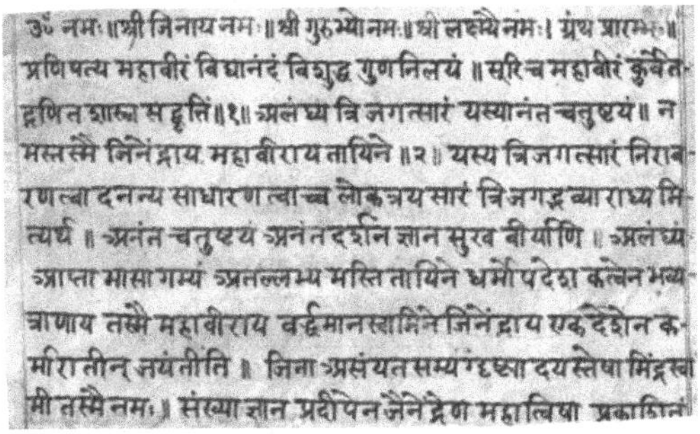
ॐ नमः ॥श्री जिनाय नमः ॥श्री गुरुभ्यो नमः॥श्री लक्ष्म्यै नमः। ग्रंथ प्रारम्भः॥
प्रणिपत्य महावीरं विद्यानंदं विशुद्ध गुणनिलयं ॥सूरिश्च महावीरं कुर्वेत-
द्गणित शास्त्र सद्वृत्तिं॥१॥ अलंघ्य त्रिजगत्सारं यस्यानंत चतुष्टयं॥ न
मस्तस्मै जिनेंद्राय महावीराय तायिने ॥२॥ यस्य त्रिजगत्सारं निराव-
रणत्वादनन्य साधारणत्वाच्च लोकत्रयसारं त्रिजगद्भव्याराध्य मि-
त्यर्थ ॥ अनंत चतुष्टयं अनंत दर्शन ज्ञान सुख वीर्याणि ॥ अलंघ्यं
आप्ताभासागम्यं अलब्ध्य मस्ति तायिने धर्मोपदेश कर्त्रेन भव्य-
त्राणाय तस्मै महावीराय वर्द्धमानस्वामिने जिनेंद्राय एकदेशेन क-
मारातीन् जयंतीति ॥ जिना असंयतसम्यग्दृष्ट्यादयस्तेषामिंद्रस्वा-
मी तस्मै नमः॥ संख्याज्ञान प्रदीपन जैनेंद्रेण महात्विषा प्रकाशितं

Uttara Chattīsī Tīkā, Commentary on *Gaṇītasāra Saṃgraha* of Mahāvīrācārya, Bhattāraka Sumatikīrti, X, 16th c. CE, (i) Ailaka Pannālāla Digambara Jaina Śāstra Bhaṃdāra, Biyavara, 331, 169, Mathematics, Saṃskṛta, Devanāgarī, (Complete), (ii) Śrī Kailāsa Sāgara Sūri Jñāna Bhaṃdāra Koba, Mss No. AKGM-13910, Folios-106, (Copy available with the present author).

After the discovery of GSS of Mahāvīrācārya and its publication with English translation by M. Rangacarya (1912) many new manuscripts have been discovered, with or without new commentaries, but there has been no attempt to edit them. The Hindi, Kannada and Telugu translations of GSS are based on the English translation. Meanwhile many objections have been raised on the English translation and editing of the work. A new translation of GSS is urgently called for incorporating all new findings and explanations.

7. *Triloka Darpaṇa* by Pt. Khadagasena (1656 CE)

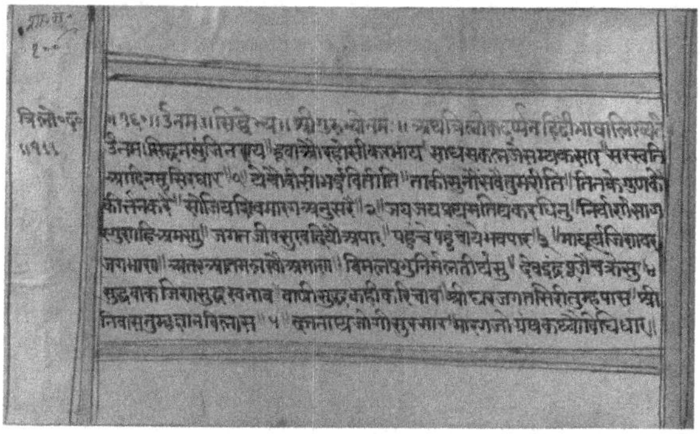

Triloka Darpaṇa, X, Pt. Khadagasena,1656 C.E, 1978 V.S. [1921 CE], (i) Amara Granthālaya, Indore, 165, 161, Structure of Cosmos, Hindi, Devanāgarī, (Complete), (ii) Ācārya Kundakunda Hastalikhita Śāstra Bhaṃdāra, Khajuraho (M.P), Mss. No 64, Folios-20. (Several other manuscripts are available in Sonagiri, Gawalior, Seoni Malwa, Indore, Sanawad (Khargone), Bina (Sagar), Shivpuri, Khargapur (Tikamgarh) etc.).

Perhaps it is the oldest Book on Jaina Cosmology written in Hindi. The style is very simple and full of pictures and tables. Its contents are the same as of *Tiloyapaṇṇatti* and *Tiloyasāra* but the presentation in Hindi makes it especially useful.

8. *Gaṇitsāthsau* of Mahimodaya (1665 CE)

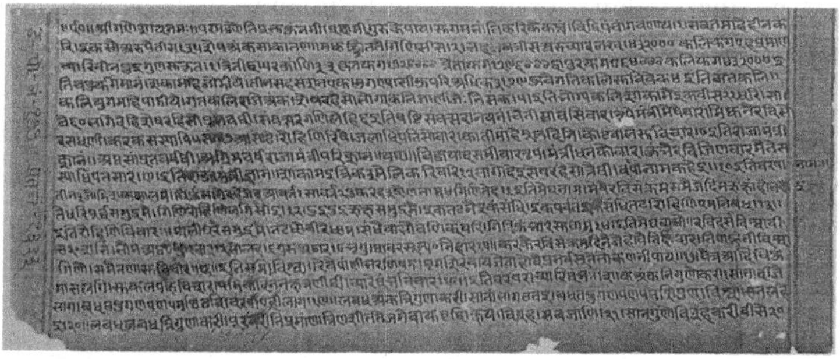

Gaṇitasāthsau, X, Mahimodaya, 1665 CE, 1694 CE, (i) Shri Mohanalal Jain, Gyan Bhaṃdāra, Gopipurā-Surat, 1333, 05,

Mathematics, Hindi/Saṃskṛt, Devanāgarī, (Complete), (ii) Abhaya Granthāgāra, Bikaner, Mss No. 5552, (iii) Vijayarāmcandrasūri Ārādhanā Bhavana, Pāchiyānipole, Ahmedabad, Mss No. 878, (iv) Śrī Sobhācandra Gyāna Bhaṃdāra, Jodhpur, Mss No. 812

This seems to be a work on mathematical astronomy. The same author has also written another book *Janmapatrīvidhi*; the present author has acquired the latter from Abhaya Granthāgāra Bikaner (Mss. No.-5548). The author has also acquired a copy of the manuscript of Gaṇīta Sāthasau which is also preserved in Abhaya Granthāgāra, Bikaner.

9. *Gaṇitasāra* of Hemarāja (17th c. CE)

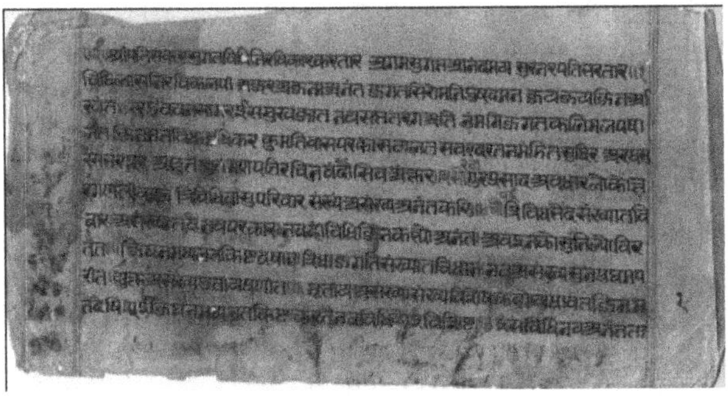

Gaṇitasāra, X, Hemarāja Godīkā, 1673 C.E approx., X, (i) Śri Jaina Vīra Pustakālaya, Ganjbasoda (Vidisha), 151, 07, Mathematics, Hindi, Devanāgarī, (Complete), (ii) Ādinātha Digambara Jain Mandira, Bundi (Raj.) Mss No. 594, Folios-5.

The present author has a copy of the manuscript from the Bundi Collection also, but the content of both the manuscripts is almost the same. The classification of numbers into countable, uncountable and infinite, which was originally given in *Tiloyapaṇṇattī* (176 CE), is discussed in this manuscript in detail in Hindi, with illustrations.

10. *Gaṇitasāra* of Muni Ānanda (Kavi) (1674 CE)

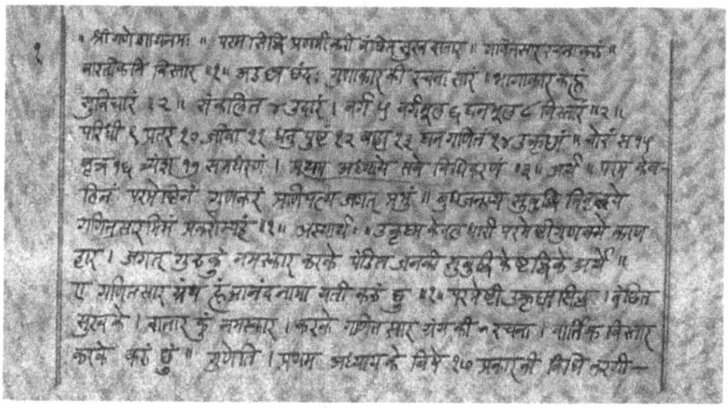

Gaṇitasāra, X, Muni Anand (Kavi), 1674 CE, 1831 V.S., (i) Sethiya Library, Bikaner, 364, 109, Mathematics, Hindi/Saṃskṛt, Devanāgarī, (Complete), (ii) Jinabhadra Sūri Jñāna Bhaṃdāra, Jaisalmer (Raj.) Mss No. 60243, (iii) Devachandra Lālabhāi Jñāna Bhaṃdāra, Surat, Mss No. 168.

This is written in a mixture of Gujarati and Hindi. Its contents have not been analyzed so far.

There are several other unpublished Jaina Mathematical manuscripts available in various libraries and manuscript repositories. The author has collected the information from the respective catalogues and data bases prepared by various organizations. Some of these catalogues were prepared about 60–70 years back, and hence availability of the manuscripts as per the record is not certain. This is especially so, since in the interim some *bhaṃdāras* (repositories) have been transferred to other places or merged with other repositories. Nevertheless, the information put together would be useful base in setting up a digital library or a physical collection of Jaina Mathematical manuscripts, which is an urgent need to ensure their conservation, for the sake of preparation for further studies. A list of these manuscripts is given in the Appendix.

12.3. Published Jaina Non-Canonical Works

Starting from 1912, to date many Jaina mathematical texts have been edited and published, some of them with translations. Here we present some information in this respect.

1. *Gaṇīta-Sāra-Saṇgraha* of Mahāvīrācārya (850 CE)

This book is written in a text book format with nine chapters. The first modern edition of it was brought out by M. Rangacarya, with an English translation and notes, published by the Madras Government, in 1912. On the basis of English translation, a Hindi translation, with notes, was prepared by L.C. Jain and it was published by Jain Saṃskriti Samrakshaka Sangh, Solapur in 1963. On the basis of same English translation, Kannada translation with notes prepared by Padmavathamma, and published by Hombuja Jain Math, Humcha, in 2000. A Telugu version followed in 2003 & 2007, published by Telugu Academy, Hyderabad.

2. *Gaṇitatilaka* Commentary on Pātīgaṇita of Śrīpati by Siṃhatilakasūri (1275 CE)

The *Gaṇitatilaka* commentary on Siṃhatilakasūri on *Pātīgaṇita* of Śripati was composed by Siṃhatilakasūri in 1275 CE. It is seen to have been influenced by the works of Śrīdhara and Mahāvīra. It was first published by Gaikawada Oriental Institute, Baroda with detailed introduction by H.R. Kapadia. Recently it was studied by Alesendra Patrochi and a version based on it is published by Routledge-London, 2019.

3. *Vyavahāra Gaṇīta* of Rājāditya (1190 C.E.)

Rājāditya is a less known mathematician of the Jaina tradition but his work is very important and is of practical use. His most well-known book is *Vyavahāra Gaṇita*. It was originally published by the Madras Government with a Kannada translation by Mariyappa Bhatt in 1955. A few years back it has been published again, with an English translation and notes by Padmavathamma *et al.* in 2013.

4. *Gaṇītasāra Kaumudī* of Ṭhakkura Pherū (1265–1330 C.E.)

The book was originally published some decades back with the other works of Ṭhakkura Pherū. A critically edited and translated version was brought out by group of Historian of Mathematics bearing a collective name 'SAKHYA' (S.R. Sarma, T. Kusuba, T. Hayashi and M. Yano), which was published by Manohar books, Delhi in 2009.

5. *Istāṃkapancaviṃsátikā* by Tejasiṃghasūri (Before 1686 CE)

This is a small book consisting of only 3 Folios written by Loṃkāgacchīya Tejasiṃghasūri in 1686 CE. It is preserved at the L.D. Institute of Indology. A copy collected by the present author was first published in Arhat Vacana [7] in Facsimile form with a short introduction. It is a book related to Astrology. An English translation of the Arhat Vacana version, together with critical notes, was published by Takao Hayashi in *Gaṇīta Bhāratī* [8].

6. *Līlāvatī* of Rājāditya (1190 CE)

This is a small book on practical arithmetic; notwithstanding its name, it has no relation with *Līlāvatī* of Bhāskarācārya (1150 CE). A translation of the book has been published together with *Vyavahāra Gaṇita* of same author (mentioned above) by Padmavathamma (2013).

7. *Kṣetra Gaṇita* of Rājāditya (1190 CE)

This is a newly discovered book by Rājāditya (1190) in the area of geometry, dealing especially with problems concerning sale and purchase of the land. It has been translated by Virupākṣa Korgal [9].

Apart from the 7 texts as above here are some more texts of which only the original version is published.

8. *Aṇgula Saptati*

This is a small book with 70 verses in Prākṛta. Its original text has been published from Cambey [10].

9–10. *Līlāvatī* and *Aṅka Prastāra* of Lālacanda (1727 C.E.)

These two small books have been published in *Shilpa Sansāra* (Magazine) Bikaner (1967) [11].

In conclusion we would like to remark that the academic world needs to take a lead in the following:

1 — Collect copies of available manuscripts from various *bhaṃdāras* (repositories) and print the facsimiles or put them on the

internet so that the interested scholars may be able to translate them and to evaluate critically their content.

2 — Catalogue the manuscripts appropriately to make the references retrievable more easily.

References

[1] *Triśatikā* (Also known as *Gaṇītasāra*) of Śrīdhara, ed. Sudhakar Dvivedi, Varanasi, 1899. Published with Sukṣemā Hindi translation of Sudyumna Acharya, New Delhi-2004.

[2] See Moodbidri mss and N. C. Shastri, Śrīdharācārya, Jaina Siddhānta Bhāskara (Arrah), **4**(1), PP. 31–42, 1947.

[3] Usha Asthana, Ācārya Śrīdhara and his Triśatikā, Doctoral thesis, Agra University, Agra, 1960, Śloka No-98 of newly prepared version.

[4] *Jyotirjñānavidhi* of Ācārya Śrīdhara, Arrah Manuscript, Folio No-4 & 5 Anupam Jain, Ācārya Śrīdhara evam unakī *Jyotirjñānavidhi*, Kundakunda Jñānapīṭha, Indore, 2012, chapter of Mss, P-59, 60.

[5] Anupam Jain, Ācārya Śrīdhara evaṃ unakī *Jyotirjñānavidhi*, Kundakunda Jñānapiṭha, Indore, 2012.

[6] Ācārya Jinasena (I), *Harivaṃśapurāṇa*, Bhāratīya Jñānapīṭha, Delhi, 10th Edition, 2006.

[7] *Arhat Vacana* (Indore), **1**(2), December 1988.

[8] *Gaṇita Bhāratī* (Delhi), **28**(1–2), PP. 129–145 2006.

[9] Śrī Kṣetra Swadi Digamabara Jain Sansthāna Math, Sonda-Sirsi (Karnataka) 2019.

[10] Atma Kamal Library, Cambay, Issue No. 3, 1928.

[11] The information provided by Late Śrī Agarchand Nahta through his letters dt. 10.3.82 and 2.6.82. I was not able to get an opportunity to see this issue of *Shilpa Sansara*.

Appendix

Some other unpublished Jaina Mathematical Texts:

These texts are preserved in different libraries. The information here is as per the catalogues of various *bhaṃdāras* (Libraries) or the databases of the certain organisations. Actual availability of the manuscripts/texts at the indicated locations has not been confirmed; even so the information would hopefully be useful in tracking desired work from the list.

1. *Gaṇividyā Paṇṇattī*

This is preserved in Bengal Asiatic Society, Collection of Oriental Manuscripts, No. 7498, as per the catalogue prepared by Pt. Kunjbihārī Nyāyabhūṣaṇa, Calcutta, 1908.

2. *Gaṇita Saṃgraha* - Yallācārya

This is reported in the Catalogue of Saṃskṛta Manuscript of Mysore and Kambey (Khambat) Library, P-318, No. 2879. Its reference is given in Jinaratna Kośa, Poona (P-98).

3. *Kṣetra Gaṇita* - Nemicanda

According to Jinaratna Kośa, Poona (P. 98) this is in the Tholiā Upāśrya Gṛaha, Ahmedabad, Pothī No. 31, Box No. 104, However no such *upāsarā* is found at the location presently. Perhaps the collection may have been shifted to another place, or merged with any other collection.

4. *Kṣetra Samāsa* - Somatilaka Sūri

According to the Catalogue of Rajasthan Oriental Research Institute, Jodhpur, prepared by Muni Jinavijaiji, 1960, Part-II, this manuscript has 32 folios. It seems to be from the 16th century. Two other Manuscripts under the name Kṣetra Samāsa Prakarāga (Prakaraṇa) and Kṣetra Samāsa Prakaraṇasāvacūri are also available in Godiji Jain Jñāna Bhaṃdāra, Pāyadhūnī-Mumbai & Jain Jñāna Bhaṃdāra, Surendranagar respectively. This information is taken from the database of Śruta Bhavana, Pune.

5. *Kṣetra Samāsa Prakaraṇa* - Śrīcandra Sūri

According to the Catalogue of Palm leaf Manuscripts in Śrī Śāntinātha Jaina Bhaṃdāra. Kambey (Khambat), Prepared by Muni Puṇyavijayaji, Gaikwad Oriental Series Publication No. 135 (1961), this is preserved as Mss No.-109 with Folios-145. A photocopy of it is available with the present author.

6. *Bṛahat Kṣetra Samāsa* Vṛatti - Siddha Sūri (Upkeśapurī)

In the Catalogue of Manuscripts in Jaina Grantha Bhaṃdāras of Jaisalmer by C.D. Dalal & L.B. Gandhi, No. 235, two copies are reported from Jaina Grantha Bhaṃdāra of Patan (Period is 1217 CE).

7. *Laghu Kṣetra Samāsa* Vṛatti - Haribhadra Sūri

According to the Catalogue of Jaisalmer Grantha Bhaṃdāra, Page No. 35 & No. 268, this is preserved in Jaisalmer.

8. *Kṣetra Samāsa-Ratna* Śekhara Sūri (14th c. CE), Pupil of Hematilaka Sūri

According to the Catalogue of Rajasthan Oriental Research Institute, Jodhpur, Page No. 2536 & Candrasāgarasūri Jñāna Bhaṃdāra, Ujjain, Mss No-1268 having Folios-17, one copy is available in each of the repositories. In the database of Śruta Bhavana, Pune, copies of Kṣetra Samāsa (Laghu) are reported, to be available in Surat, Kolkata, Ladnun and Ujjain, with the authorship of Ratna Śekharasūri. The present author has acquired one copy from Mahāvīra Ārādhana Kendra-Koba.

9. *Kṣetra Samāsa* by Siṃhatilaka Sūri (Pupil of Somaprabha Sūri)

According to the database of Śruta Bhavana, Pune, 5 copies of this title are available in Gujarati Jaina Śvetāmbara Tapāgaccha Jñāna Bhaṃdāra, Kening street, Kolkata No 2532–2536.

10. *Kṣetra Samāsa Bṛahada* by Jinabhadragaṇi with Commentary of Malayagiri

According to database of Śruta Bhavana, Pune, Siddhimegha Manoharasūri Śāstra Bhaṃdāra, Part-1 Saṃvegī Upāsarā, Hajipatel Chowk, Ahmedabad, Box/Sr/Mss/195/3978 & 3979 Folio 108 & 150. Several other copies are reported in Surendranagar, Raghanpur and Koba with the authorship of Jinabhadragaṇi.

11. *Uttara Chattīsī Tīkā* - Śrīdhara etc.

As per listed information, one copy of this is available in Digambara Jaina temple Balātkāragaṇa, Kāranjā, Bastā No-13. However, in the author's opinion it is perhaps an incomplete version of Manuscript of GSS of Mahāvīrācārya, as Śrīdharācārya is not known to have written any such book.

12. *Gaṇita Śāstra* (Commentary) - Guṇabhadra

As per information in the Digambara Jaina temple, Balātkāragaṇa Kāranjā, Vesthan No-13, Mss No-64, is a copy of *Gaṇita Śāstra*. Guṇabhadra (9–10th C.) is a Digambara Jaina ācārya who has written Uttarapurāṇa but not any mathematical work is reported so far.

13. *Gaṇita Vilāsa* (*Gaṇita Sāra*) - Candrama (1650 C.E.)

As mentioned in the Kannada work Prāntīya Tādapatrīya Grantha Sūcī, Pt. K. Bhujabali Shastri, a few copies are available in Jaina Math-Moodabidri, Mss No-160, Folio-29, Jaina Bhavan, Moodabidri, Mss No-89, Folios-32, is complete & Mss No-216, Folio-10, is incomplete.

14. *Gaṇita Śāstra* - Rājāditya

As mentioned in the Bibliography of Saṃskṛta works on Astronomy and Mathematics by S.N. Sen *et al.* P-207, one copy is available in the Oriental Manuscript Library of Fort St. George College, Madras, Folios-25. This Book is not mentioned in the list of works of Rājāditya hence it needs to be checked whether it is a work of his that was unknown so far to belongs to another author (and wrongly catalogued). At present we have information of 7 books by Rājāditya.

15. *Gaṇita Saṃgraha* - Rājāditya

As per Kannada Prāntīya Tādpatrīya Granth Sūchī one copy is in Jain Matha, Moodabidrī, Mss No-590, Folios-9. No such work is listed so far as work of Rājāditya. It should be examined.

16. *Gaṇita Vilāsa* - Rājāditya

In the same way in Vaikanatakāra basadi, Moodabidrī we have Mss. No-7, Folio-16. Jain Matha, Kārakala, Mss No-54 (three copies) Folios-19,15 & 60.

17. ***Muttina Cippana Sūtragalu*** **- Rājāditya**

The copy of this newly reported manuscript is preserved with Shri Virupakṣa Korgal. The original copy is preserved in Mysore University Collection and Madras University Chennai (Mss No. KA 132/2) collection, as reported by Virupakṣha Korgal. It will be published shortly, with a translation.

18. ***Pudgala Bhṃga & Vṛatti*** **- Naya Vijaya Gaṇi**

We find it reported in Bhandārakar Oriental Research Institute, Pune. Mss. No-215, Vijayaramcandrasuri Ārādhanā Bhawana, Pāchīyānī Pole, Ahmedabad, Catalogue Page No-147, Mss No-878, No. of Folio-12, Hans Vijai Śāstra Saṃgraha, Ghadiyali Pole Vadodara, Catalogue Page No 56A, Box/177, Mss 2324, Folios-6.

19. ***Gaṇividyā Prakīraṇaka*** **(Prākṛta) - Sthavira**

Two copies of this are reported in the Catalogue of Rajasthan Oriental Research Institute, Bikaner, Vol-13, Mss No-13095, Folio-4 & Jain Mahājan Gyan Bhandar, Kadaya-Kaccha Box-65, Mss-296.

20. ***Gaṇividyā*** **(Prākṛta)**

One copy is reported in Tapovana Citakoṣa, Navasarī, Mss. No-678, Folios-5.

21. ***Gaṇividyā Payanno*** **(Prākṛta)**

One copy of it is reported in Mohanlal Jain Śvetāmbara Jñāna Bhaṃdāra, Gopipura-Surat, Pothi No-5, Mss No-38, Folios-6.

22. ***Gaṇivijjā Painnā***

In the catalogue of Asiatic Society Government Collection Vol. XIII, 4310/III, Catalogue-80. Gaṇi Vidyā Payanno & Gaṇivijjā Painnā may be the same.

23. ***Kṣetrasamāsāvacūri*** **- Guṇaratna Sūri**

This is mentioned in Bibliography of Saṃskṛta works on Astronomy and Mathematics, S.N. Sen *et al.*, P-86.

24. *Kṣetrasamāsa Vivaraṇa* - Jaishekhara Sūri

This is reported in Gujarātī Jain Śvetāmbara Murtipūjaka Tapāgaccha Jñāna Bhaṃdāra, Kenning Street, Kolkata, Mss. No Da/Pra-9/428, Folio-40 and Saṃvegi Jain Upāśraya, Masjid Chowk, Badwada (Gujarat), Mss. Da/Pra/9/426 Folio-40 etc.

25. *Bhāṣā Līlāvatī* by Tejasingh

This is mentioned in the Catalogue of Oriental Institute of Vadodara, Vol-4, Page No. 552, 553, Folio-10.

26. *Bhāṣā Līlāvatī* by Ānandamuni

This is mentioned in the Catalogue of Oriental Institute of Vadodara, Vol-4, Page No. 552, Folio-11.

27. *Bhāṣā Līlāvatī Gaṇapata*, Lālendra Jain (Kavi)

This is mentioned in the Catalogue of Hemacandrācārya Pāthaśālā, Pālītānā, Page-30, Mss. 471, Folio-24.

28. *Gaṇitasāra Saṃgraha Chhatīsītīka* with Kannada commentary

A copy of this Chattīsīṭīkā is available in NIPSAR-Sravanbelgola. It is in *Kānarī* Script and the Institute is planning to translate and publish it. It is said to be a work of Mahāvīrācārya, or based on one such.

29. *Gaṇitasāra* or *Triśati Gaṇitasāra* of Śrīdharācārya

The published edition of Traśatikā of Śrīdhara is incomplete, therefore these two manuscripts are important for preparation of a new edition of Gaṇitasāra. The details are as follows — Mohanlal Library-Mumbai, Catalogue, Page No-6, Mss. No-156, Folio-17, Jinabhadrasūri Jñāna Bhaṃdāra, Jaisalmer, Folio-34.

30. *Gaṇita Koṣṭhaka*

This is reported in Jain Matha, Kārkala, Mss No-54.

31–33. *Gaṇita Kaumudī, Gaṇita Līlā* (Pt. Bhāsakāra), *Gaṇita Nāmamālā*

These are reported in Jain Granth Bhaṃdāra, Amer, as per List of K.C. Kasliwal, Jaipur.

34. *Caturaṇga Lekhana Tathā Taṇdula Sthāpana Krama*

It is reported in Kannada Prāntīya Tādapatrīya Grantha Sūcī, Pt. K. Bhujabali Śāstrī, Page-169, Mss. No-285.

35. Apart from these, present author has collected the following manuscripts also from various places

1. *Aṁka Laharī*
2. *Janmapatrīvidhi*
3. *Līlāvatī* by Lālacandra
4. *Līlāvatī* by Mathurānātha
5. *Bhāṣā Līlāvatī* by anonymous author
6. *Ghanagaṇita Saṃgraha*
7. *Laghu Saṃgrahaṇī*
8. *Saṃgrahaṇī Sūtra* etc.